IFIP Advances in Information and Communication Technology 469

Editor-in-Chief

Kai Rannenberg, Goethe University Frankfurt, Germany

IFIP – The International Federation for Information Processing

IFIP was founded in 1960 under the auspices of UNESCO, following the first World Computer Congress held in Paris the previous year. A federation for societies working in information processing, IFIP's aim is two-fold: to support information processing in the countries of its members and to encourage technology transfer to developing nations. As its mission statement clearly states:

> IFIP is the global non-profit federation of societies of ICT professionals that aims at achieving a worldwide professional and socially responsible development and application of information and communication technologies.

IFIP is a non-profit-making organization, run almost solely by 2500 volunteers. It operates through a number of technical committees and working groups, which organize events and publications. IFIP's events range from large international open conferences to working conferences and local seminars.

The flagship event is the IFIP World Computer Congress, at which both invited and contributed papers are presented. Contributed papers are rigorously refereed and the rejection rate is high.

As with the Congress, participation in the open conferences is open to all and papers may be invited or submitted. Again, submitted papers are stringently refereed.

The working conferences are structured differently. They are usually run by a working group and attendance is generally smaller and occasionally by invitation only. Their purpose is to create an atmosphere conducive to innovation and development. Refereeing is also rigorous and papers are subjected to extensive group discussion.

Publications arising from IFIP events vary. The papers presented at the IFIP World Computer Congress and at open conferences are published as conference proceedings, while the results of the working conferences are often published as collections of selected and edited papers.

IFIP distinguishes three types of institutional membership: Country Representative Members, Members at Large, and Associate Members. The type of organization that can apply for membership is a wide variety and includes national or international societies of individual computer scientists/ICT professionals, associations or federations of such societies, government institutions/government related organizations, national or international research institutes or consortia, universities, academies of sciences, companies, national or international associations or federations of companies.

More information about this series at http://www.springer.com/series/6102

Eunika Mercier-Laurent · Mieczysław Lech Owoc
Danielle Boulanger (Eds.)

Artificial Intelligence for Knowledge Management

Second IFIP WG 12.6 International Workshop, AI4KM 2014
Warsaw, Poland, September 7–10, 2014
Revised Selected Papers

 Springer

Editors
Eunika Mercier-Laurent
Jean Moulin University Lyon 3
Lyon
France

Danielle Boulanger
Jean Moulin University Lyon 3
Lyon
France

Mieczysław Lech Owoc
Wroclaw University of Economics
Wroclaw
Poland

ISSN 1868-4238   ISSN 1868-422X   (electronic)
IFIP Advances in Information and Communication Technology
ISBN 978-3-319-28867-3   ISBN 978-3-319-28868-0   (eBook)
DOI 10.1007/978-3-319-28868-0

Library of Congress Control Number: 2015960815

Printed on acid-free paper

This Springer imprint is published by SpringerNature
The registered company is Springer International Publishing AG Switzerland

Preface

Knowledge is still one of intangible capitals that influence the performance of organizations and their capacity to innovate. From the very beginning the Knowledge Management (KM) movement initiated in the last century has proposed various approaches focused on supporting enterprises including nonprofit organizations. Traditionally, knowledge gathering, knowledge modeling, and knowledge applications (i.e., knowledge processing) are the main topics essential in KM endeavors. It seems reasonable to join this stream with methods and solutions offered by both symbolic and computational intelligence; we often need to combine both for the best results.

After the first AI4KM (Artificial Intelligence for Knowledge Management) organized by IFIP (International Federation for Information Processing) group TC12.6 (Knowledge Management) in partnership with ECAI (European Conference on Artificial Intelligence) 2012, the second workshop was held during the Federated Conferences on Computer Science and Information Systems (Fedcsis) 2014 in conjunction with the Knowledge Acquisition and Management Conference (KAM).

The main objective of this conjunction was to gather both researchers and practitioners to discuss methodological, technical, and organizational aspects of AI used for knowledge management and to share the feedback on KM applications using AI, especially for business. The main stream of this event was particularly devoted to selected aspects of collaborative human-machine intelligence.

We would like to thank the members of the Program Committee, who reviewed the papers and helped put together an interesting program in Warsaw. We would also like to thank the invited speaker and authors. Finally, our thanks go to the local Organizing Committee and all the supporting institutions and organizations.

This volume offers a selection of improved papers presented during the workshop and includes one invited paper. After the presentation, the authors were asked to extend their proposals by highlighting their original thoughts. The selection focused on new contributions to the KM field and innovative aspects. An extended Program Committee then evaluated the final versions of the proposals, leading to these proceedings.

The proceedings begin with the invited paper:
"A Sign-Based Management Methodology for Co-designing Educational E-services in Living Labs," by Noël Conruyt, Véronique Sébastien, Olivier Sébastien, Didier Sébastien, and David Grosser.

The rest of the papers are organized according to the four tracks at the workshop:

- Tools and Methods for Knowledge Acquisition

The first paper, "Role of Data Warehouse as a Source of Knowledge Acquisition in Decision-Making. An Empirical Study," responds to the scarcity of empirical studies examining the data warehousing success within an integrative model.

The next contribution "Knowledge Extraction from Professional E-mails" deals with volatile knowledge when professional actors interact together and tackle problems in order to realize projects.

- Models and Functioning of Knowledge Management:

"Challenges for Knowledge Management in the Context of IT Global Sourcing Models Implementation" covers the determination and management of the most important risks related to information sharing in IT sourcing with particular attention to various cloud computing services on offer.

The article "How Should Digital Humanities Pioneers Manage Their Data Privacy Challenges" focuses on the concept of "privacy by design" to address the right of data protection and to guarantee the movements of personal data exchanged between business stakeholders and member states.

The innovative part of the research presented in "Usability of Knowledge Portals for Exclusives in Local Governments" is devoted to the concept of knowledge portals covering architectures and examples of supported tasks.

The next paper, "Knowledge Management in Distributed Agile Software Development Projects: Techniques, Strategies and Challenges," investigates knowledge-sharing techniques and strategies applied by practitioners in the context of distributed agile projects.

- Techniques of Artificial Intelligence Supporting Knowledge Management

The main objective of the study "Actuator Fault Diagnosis Using Single and Meta-Classification Strategies" is to compare either single or meta-classification strategies that can be successfully used as a reasoning means in diagnostic-aided expert system. The authors propose a new approach for searching proper values of relevant parameters of classifiers used for fault diagnosis.

The article "Intelligent Association Rules for Innovative SME Collaboration" provides a pre-analysis of the path of successful SME alliances leading to improvements in innovative power. The implication of the study is generic enough to help any SME or research organization or large business to reduce risks in future alliances.

- Components of Knowledge Flow

The contribution "Managing Intellectual Capital in Knowledge Economy" presents an overview of experiences and research works in applying artificial intelligence approaches and techniques for intellectual capital management. The article also describes a method and tools to treat this wealth differently and to activate and stimulate a discussion on the role of this capital in knowledge economy and in innovation ecosystems.

The papers cover essential subjects in collaborative human–machine intelligence and reflect research performed at different academic centers.

We hope you will enjoy reading these papers.

October 2015

Eunika Mercier-Laurent
Mieczysław Lech Owoc
Danielle Boulanger

Organization

Co-editors

Eunika Mercier-Laurent Jean Moulin University Lyon 3, France
Mieczysław Owoc Wroclaw University of Economics, Poland
Danielle Boulanger Jean Moulin University Lyon 3, France

Program Committee

Danielle Boulanger Jean Moulin University Lyon 3, France
Eunika Mercier-Laurent Jean Moulin University Lyon 3, France
Nada Matta Troyes Technical University, France
Mieczyslaw Lech Owoc Wroclaw University of Economics, Poland
Anne Dourgnon EDF Research Center, France
Otthein Herzog Jacobs University, Bremen, Germany
Daniel O'Leary USC Marshall School of Business, USA
Antoni Ligeza University of Science and Technology, Krakow, Poland
Helena Lindskog Linköping University, Sweden
Gülgün Kayakutlu Istanbul Technical University, Turkey
Knut Hinkelmann University of Applied Sciences and Arts, Switzerland
Vincent Ribière Institute for Knowledge and Innovation, Bangkok, Thailand
Jean Rohmer Pole Leonard de Vinci, France
Frédérique Segond Objet Direct, Grenoble, France
Eric Tsui Hong Kong Polytechnic University, SAR China

Local Organizing Committee

Fedcsis 2014, Warsaw, Poland – Marcin Paprzycki, Maria Ghanza, Leszek Maciaszek

Contents

Sign Management for the Future of e-Education: Examples of Collaborative
e-Services in a Living Lab (Invited Paper)   1
   Noël Conruyt, Véronique Sébastien, Olivier Sébastien, Didier Sébastien,
   and David Grosser

Tools and Methods for Knowledge Acquisition

The Role of Data Warehouse as a Source of Knowledge Acquisition
in Decision-Making. An Empirical Study   21
   Moh'd Alsqour and Mieczysław L. Owoc

Knowledge Extraction from Professional E-mails   43
   Nada Matta, Hassan Atifi, and François Rauscher

Models and Functioning of Knowledge Management

Challenges for Knowledge Management in the Context of IT Global
Sourcing Models Implementation   58
   Kazimierz Perechuda and Małgorzata Sobińska

How Should Digital Humanities Pioneers Manage Their Data
Privacy Challenges?   75
   Francis Rousseaux and Pierre Saurel

Usability of Knowledge Portals for Exclusives in Local Governments   92
   Krzysztof Hauke, Mieczysław L. Owoc, and Maciej Pondel

Knowledge Management in Distributed Agile Software Development
Projects   107
   Mohammad Abdur Razzak, Touhid Bhuiyan, and Rajib Ahmed

Techniques of Artificial Intelligence Supporting Knowledge Management

Actuator Fault Diagnosis Using Single and Meta-Classification Strategies . . .   132
   Mateusz Kalisch, Piotr Przystałka, and Anna Timofiejczuk

Intelligent Association Rules for Innovative SME Collaboration   150
   Gulgun Kayakutlu, Irem Duzdar, Eunika Mercier-Laurent,
   and Bahar Sennaroglu

Components of Knowledge Flow

Managing Intellectual Capital in Knowledge Economy   165
  Eunika Mercier-Laurent

Author Index   181

Sign Management for the Future of e-Education: Examples of Collaborative e-Services in a Living Lab (Invited Paper)

Noël Conruyt[✉], Véronique Sébastien, Olivier Sébastien,
Didier Sébastien, and David Grosser

LIM, EA 25-25, University of Reunion Island, 97490 Sainte-Clotilde, France
{noel.conruyt, veronique.sebastien, olivier.sebastien,
didier.sebastien, david.grosser}@univ-reunion.fr

Abstract. From a European technological and industrial perspective of the 20th century, Knowledge Management (KM) was viewed as the next step towards reaching a smart knowledge-based economy. But today, in the 21th century of big data and fast moving information, we argue that KM is not enough for reaching a qualitative human-based post-industrial society. We need a broader view in order to understand user needs and respond to their personal desires. In this endeavor, Living labs are a good way to reposition creative people at the center of technologies. But we need also methodologies and tools to accompany the transition from a competitive economy to a more sustainable society. We experimented this move at University of Reunion Island in the domain of e-education. We conceived a new paradigm called Sign Management (SM) for enhancing content producers with multimedia tools on a Creativity Platform. A methodology for co-designing educational e-services was applied in both natural (biodiversity) and cultural (music) domains in order that linear knowledge transmission lets place to an iterative know-how sharing approach between teachers and learners. This sign-based methodology serves as a condition for opening the era of Semiotic Web (Web of Signs) over Semantic Web (Web of Things). The objective is to co-create qualitative educational e-services with people based on a more natural/artificial and intelligent approach in the frame of Living labs.

Keywords: Sign management · Semiotic web · e-service · e-education · Living lab · Creativity platform

1 Introduction

The Lisbon strategy of Europe [1] tried to make the EU "the most competitive and dynamic knowledge-based economy in the world capable of sustainable economic growth with more and better jobs and greater social cohesion", by 2010. But it failed because of a techno-centric approach of innovation principally focused on economic growth. It forgets the role of human beings in the development of a knowledge society that takes into account social and environmental awareness. A quick count at the report from the High Level Group [2] of the word *competition* on one hand (12 occurrences)

© IFIP International Federation for Information Processing 2015
E. Mercier-Laurent et al. (Eds.): AI4KM 2014, IFIP AICT 469, pp. 1–20, 2015.
DOI: 10.1007/978-3-319-28868-0_1

and *cooperation* or *collaboration* on the other hand (4 occurrences) is enlightening. It shows the direction followed by EU in the new millennium for achieving the knowledge economy in a techno-centered industrial understanding of sustainability, moving first to economy (growth and jobs), then to social and environment resolution. The result is the one that is described by Luycks Ghisi [3] on knowledge society: the negative scenario of engineering the human mind and make a green washing of our environment to mask the aim of doing business as usual!

The positive scenario is another open post-industrial vision, i.e. reaching a sign-based society. It is not only a matter of technological, economical, or social awareness; it is also grounded in individual, human, environmental and cultural values. Psychological, ethical, biological and emotional assets are indeed drivers of the future of a knowledge society in a perspective of sustainable development of services co-designed with people. Sustainability applied to open innovation 2.0 needs to give empowerment to users for them to drive new products and services that fit their real and useful needs. For Curley and Salmelin, it is the only way for their adoption [4].

In this paper, we will apply this vision of a sign-based society to the problem of open education 2.0 with Information and Communication Technologies (ICT). We will first explain the role of ICT in this change of paradigm for the Future Internet and Web. We will then analyze how Living Labs help us to manage this enlarged definition of ICT. Then, we will introduce our theory of sign management that is based on three levels of capacity: personal and phenomenal volition (Biological Semiotics), relational and verbal action (Activity Theory) and formal and written cognition (Signification Theory). This theory leads to the new paradigm of Semiotic Web that encompasses other Web dimensions, i.e. Service, Social, Semantic and Immersive Web. This theory has been applied in our University of Reunion Island Living Lab for Teaching and Learning how to manage biodiversity and music heritage. We will show two examples of this new approach for innovating with users in educational contexts. These examples of collaborative e-services in a Living Lab emphasize the need of sharing know-how rather than transmit knowledge for the future of e-education.

2   ICT in the Future of a Knowledge-Based Society

For the objective of reaching the future of a knowledge-based society (that we call a sign-based society), we need to understand the meaning of words that are currently used in our digital world (ICT, Internet, Web, Services, objects, etc.). For example, Fig. 1 shows that ICT must not only be considered as technological tangible *containers* (Technologies), but also convey human *contents* (data, information, and knowledge). These intangible assets are communicated between users (Information and Communication) by the mean of network products (virtual platforms), electronic services (e-services) and Web applications in the clouds.

At the upper level of ICT, there is the Web in real contexts with people. The Web is a service of Internet (as the email or the ftp) to connect producers and consumers of multimedia contents, in order that Internet infrastructures meet user needs [5]. Co-designing e-services with users is an iterative communication process that is central between Internet technology and Web usage. These collaboration principles have been adopted since 2006 by the European Network of Living Labs (ENoLL) and are developed in the frame of corresponding literature [6]. The changing world philosophy is that in a sustainable society, the political leaders must concentrate on the demand of subjects rather than on the offer of objects. The innovation path is to transform a quantitative strategy of things delivered massively on the Internet (IoTs) by a qualitative strategy of signs on the Web (WoSs) that are more personalized in order to be more meaningful for individuals and communities.

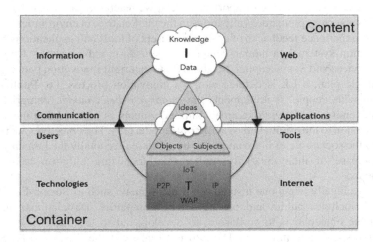

Fig. 1. ICT are not only a technological matter but also cloud services that matter!

Indeed, technologies are means, not objectives. As says M. Giget, "Very few people fall in love of a Wireless Access Protocol or Internet of Things!" [7]. So, innovation relies also in co-creating products and services that end-users would like and use. As an example in e-education, some of them are researchers and specialists of a domain, such as in natural sciences or in arts. They acquired knowledge through their long practice of domain objects (species, pieces), and are recognized as experts by other practitioners. The amateurs would love to be educated by masters, but the physical distance is an obstacle for them to be taught. Indeed, the best way to learn a domain is to get a course with a professor. With ICT, finding solutions that recreate the conditions of a real course on the Web can tackle this distance educational problem and respond to real user needs.

3 Setting up ICT Living Labs for Education

ICT Living Labs position themselves at the center of the process of convergence of containers and contents for making tools and applications. It can be represented by the triangle with the clouds' birth in Fig. 1. They allow users (subjects) to communicate ideas between them about things (objects) in order to co-design new e-services on the

Web. Arrows show that there is an iterative process to set up for building content applications that depend on container generator tools. Living Labs can help us to manage this iterative process of going from ideas to services and products.

For us, a Living Lab (LL) is both a real and virtual environment for user-centered innovation, based on the observation of every-day user practice and experience for solving problems, but also based on their active participation, with an approach that facilitates their influence in the open and distributed innovation process (participatory design). It engages all concerned partners in the real-life contexts, and aims to create sustainable usage values [8]. As important effects, we reach relevant knowledge on the context of use, such as premature validations in the market, tries in environments familiar to the users, experience feedbacks of co-creative users of tools and applications [9]. This new innovation system is a human-centered ecosystem based on expert Knowledge, Business services and Social capital, i.e. the KBS Concurrent Innovation paradigm [10]. On a strategic plan, a LL is defined as a 4P innovation process, i.e. Public-Private Partnership with People. It implements the *quadruple helix model*, comprising four strands: university, industry, government and individuals as users adding the society's fourth dimension to the original innovation triple model [11]. The objectives are political, focused on the social role of innovation, i.e. trying to realize totally the human potential by the increase of their creativity, rather than focalizing only on technological developments.

More specifically when one instantiates it in a domain such as Social>Education, a Living Lab includes public and private actors, companies, associations, individual actors, whose objective is to co-design, to develop and to test life-size educational services, tools and new practices. The aim is to take out the Research>Teaching of laboratories to make it come down in the daily life, often by having a strategic view on the potential uses of tools and applications for doing Business. All this takes place in cooperation between local authorities, companies, research laboratories, as well as potential users, via helping platforms for designing innovative e-services and analyzes of their usages. It is a question of favoring the culture of open innovation 2.0, sharing networks, and involving the users from the beginning of the conception.

The European Institute of Innovation and Technology (EIT) was created in 2008 as an independent body of the European Union to bring together leading higher education institutions, research labs and companies to form Knowledge Innovation Communities (KICs) that develop innovative products and services, start new companies, and train a new generation of entrepreneurs. Their mission is described in [12]. One of the KICs is devoted to ICT [13]. Figure 2 shows the Knowledge Triangle of EIT ICT Labs (now called EIT Digital) that implements the societal role of universities at the center of the European vision in a Knowledge-based society [14, 15].

In their vision, Living Labs are positioned in the educational part of the triangle for making experimentations and getting experience. The educational action process is described in [16] for teaching and learning in the knowledge triangle:

The knowledge triangle has so far mostly been presented as a theoretical concept and political desideratum over changes needed in Europe. This theoretical model now must be transformed into a model of action, an everyday working model for the people involved. One way is to create an enquiry based process around the three nodes of the triangle. Three questions need to be in the mind of everyone at all levels in the system and in all planning and performing: (1) What are the best ways of linking research to education? (2) What are the best ways of teaching creativity, innovation and entrepreneurship? and (3) How can optimal conditions be created for entrepreneurs and innovators to bring their knowledge and experience back into research?

Fig. 2. The knowledge triangle

But in this process, they don't figure out the role of ICT in the triangle model. The knowledge triangle is depicted as a static descriptive view between Research, Education, and Business. In our view, it could be enhanced dynamically with ICT by a new representation of Signs and a signification process (see the next section). For understanding it, Fig. 3 shows that the triangle must be seen from the above and not from one side. The new Sign representation that we propose for integration of ICT is a 3D tetrahedron model. The Living Lab is a Co-location Center that stands at the center of the tetrahedron for communication purpose and ICT is fueling the convergence of Research, Education and Business with a new signification process.

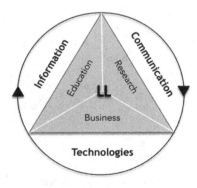

Fig. 3. The living lab tetrahedron (seen from above)

In our view, we think that we must shift from a flat 2D triangle representation of Knowledge with its linear transmission (from Research to Education to Business), to a circular 3D tetrahedron representation of Signs (Data, Information, Knowledge, Subject) and their dynamic signification process (called semiosis) for sharing them. We will explain now this new iterative sign management process for the future of knowledge management with ICT.

4 Theory

Sign Management emphasizes the engineering and use of data, information and knowledge from the viewpoint of an interpreter (Subject).

Representation of Signs. The concept of Sign is derived from the pragmatic and *triadic* Peirce's theory of semiotics with a Sign's correspondence of the Subject to its Object. From this philosophical viewpoint, a Sign, or representamen, is something that stands to somebody for something in some respect or capacity [17]. From our computer science analysis, Data (Object) is the content of the Sign (something), Information, a multi-layered concept with Latin roots ('informatio' = to give a form) is its form, and Knowledge is its sense or meaning, i.e. no-thing. The notion of Sign is then more central than knowledge for our purpose of designing e-services.

Moreover, as we want to compare different sign representations coming from different interpreters, we designed a *fouradic* representation of signs that can be communicated in a personalized way, i.e. through different interpretations of the same objet.

In Fig. 4, we define the tetrahedral representation of a Sign as the interpretation of an Object by a Subject at a given time and place, which takes into account its content (Data, facts, events), its form (Information), and its sense or meaning (Knowledge).

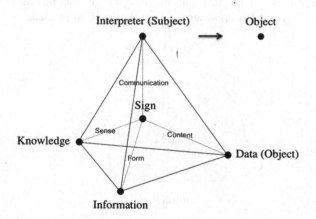

Fig. 4. The tetrahedron of the sign (static representation)

This new representation of signs enhances the subjective aspect of knowledge that can be managed with ICT, because some interpreters are better than others for solving problems: their experience of specialist can be described and shown with multimedia contents in order to learn their good practices or know-how. Managing knowledge in general is not possible without managing personal know-how that gives birth to it. So, the problem is to wonder how to construct objective knowledge with ICT from different subject interpretations, which is the aim of science?

Signification/Construction of Signs. Knowledge is the ultimate goal of Science, but this target is difficult to reach without taking into account humans who generate it in a constructivist manner!

In Fig. 5, we introduce Signification, i.e. the continuous process of using Signs in human thinking for acquiring Objects interpreted by Subjects. This sign construction process or Semiosis takes the different components of the Sign in a certain order to make a decision: first comes the Subject or Interpreter who is receptive to his milieu or "Umwelt" [18], and who cares about Information to act in a certain direction (volition, conation). Then occurs the searched Data (Object) to position himself in space and time (action). Next, Knowledge is activated in his memory to compare the actual situation with his past experiences and make a hypothesis for taking a decision (cognition).

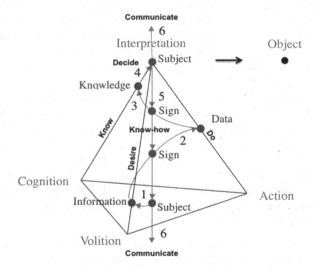

Fig. 5. The signification process (dynamic)

The signification or the building of the sign communicates the process iteratively in a reflexive way (in order to memorize new knowledge) or communicates the resulting interpretation as information to its environment (by exteriorization).

Semiosis is similar to the working principle of inference engine that was modeled in expert systems: the evaluation-execution cycle [19]. The difference is that Signification integrates the Subject in the process, and this integration is therefore more meaningful to humans than to machines. The Subject operates on Signs in two phases: reflection and action. These phases are inter-linked in a reflexive cycle with a semiotic spiral shape including six moments: (1) to desire, (2) to do, (3) to know, (4) to interpret, (5) to know-how for oneself, (6) to communicate to others (Fig. 5). The semiosis spiral is included in the tetrahedron of the Sign.

Consequently, Signification is the key psychological process that makes sense for practicing usage based research and development with people by communicating data, information and knowledge. Sign management is based on representation AND signification of Signs at three levels of capacity: personal and phenomenal volition (Uexküll Biological Semiotics Theory), relational and verbal action (Engeström Activity Theory) and formal and written cognition (Peirce Signification Theory).

Semiotic Web. When an organism or an individual seeks for something, his attitude is to pay attention to events of his environment that go in the sense (direction) of what he searches. The primary intention of a microorganism such as bacteria is "good sense": it wants to capture information from the milieu to develop itself and stay alive [20].

Human development follows the same schema of self-organized living systems at more complex levels than these physiological and safety needs. They are those that have been defined in the hierarchy of fundamental individual needs: love, belonging, esteem, self-actualization [21].

As a consequence, we hold that before being able to make "true sense", i.e. adopt a scientific rationale, the objective of individuals is to respond to psychological needs (desire, pleasure, identity, etc.). This theory of human motivation is a natural and cultural hypothesis, which is corroborated by Umwelt [18], Activity [22] and Semiotic [17] pragmatic theories. These life and logical sciences are components of the Biosemiotics interdisciplinary research [23], which was introduced before the advent of Internet as the "Semiotic Web" [24].

By comparison, Semantic Web is the *dyadic* combination of form and sense of the linguistic Sign [25], taken as a signifier (form) and signified (sense). It is rational. Semiotic Web is more generic and living. It complements the Semantic Web (form and sense) with the referents (content) that are observed data (interpretations) geo-referenced in a 3D information world (Immersive Web) as Web Services by subjects pertaining to communities of practice (Social Web 2.0). This makes our Sign management ecosystem a *tetrahedron* model (Fig. 6) that is more involved in concrete life with end-users on a specific territory such as Reunion Island.

The Web of Signs combines:

1. The Web of *Data* and *Objects*, i.e. the flow of raw and digital contents produced by specialists (teachers) and transmitted by engineers in databases and knowledge bases in the frame of an Information System (one-way flow), but progressively becoming interoperable through Web services with other Information Systems,
2. The Web of *Subjects*, i.e. a bidirectional communication platform between users (teachers and learners) using different e-services within a community of practice to exchange interpretations of data and objects, and negotiate their value,
3. The Web of *Information* that is geo localized in attractive virtual worlds representing the real landscape (metaverses), and accessible at any time, anywhere, on any devices (mobiquity).
4. The Web of *Knowledge* for machines to communicate logically on the basis of a formal, open and semantic representation of data and objects.

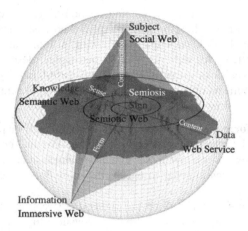

Fig. 6. The four dimensions of semiotic web: service, social, semantic and immersive web

The Semiotic Web is for us the future of the Semantic Web: it is the Web of Signs rather than the Web of Things. The Semiotic Web is more general than the Semantic Web because it allows managing Signs (Data, Information, Knowledge) from a subjective viewpoint and not only Knowledge from an objective angle. Signification is the kernel of Semiotic Web although Representation is at the root of Semantic Web. Both are necessary to co-design e-services in the future Web, but from our experience, co-designing e-services with users needs to manage signs on a Creativity Platform for building e-services that match lead-users needs.

5 Research Design

At the University of Reunion Island, we have investigated these different dimensions (phenomenal, relational and formal) that are converging to form what we call the Semiotic Web. As the World is an Island and as Reunion Island is a small world, we designed our Living Lab as a small laboratory for Teaching and Learning Sciences and Arts by Playing [26]. Indeed, *edutainment* is one of the pillars of the future Web [27]. With game-based learning, we consider that we can play seriously to better know our environment and then better protect it.

For making e-services, the first step is to generate ideas. This ideation process should be attractive for motivating some researchers to drive projects. Living Labs are those desirable innovation ecosystems (breeding grounds) for enhancing research and go to the market, i.e. make ideas become alive (emergence of ideas) towards products and services. The second step is then to co-design mock-ups and prototypes, and experiment them in a physical meeting place called the Creativity Platform (see below). Co-creation in communities of researchers, entrepreneurs and users is also a significant characteristic of open and social innovation that is part of the DNA of Living Labs for bringing trust between co-designers. The third step is to formalize a solution that fits user needs. It applies abductive, inductive and deductive science cognition principles and generates implicit knowledge (for self) or explicit knowledge (for others). Semantic knowledge is captured in knowledge bases with a tool called IKBS (Iterative Knowledge Base System) for defining written ontologies, describing use cases and making qualitative decisions with induction and case-based reasoning. Semiotic annotations with multimedia objects are proposed in order to illustrate the terms that are used in the ontology and in the cases.

The Creativity Platform. For the purpose of co-designing such a product/service with ICT, the Creativity Platform is the co-working, co-learning and communication space for researchers and developers, businesses and users, aimed at collectively defining the characteristics of e-services in order to ensure the most direct match between expectations and use.

In its technical form (see Fig. 7 on the right), the Creativity Platform includes a multimedia platform as the one that we find in television studios, but also includes a physical and virtual place to discuss ideas and projects, make models and prototypes, and experiment them in a synchronous (focus group) or asynchronous (video forum on the Web) way.

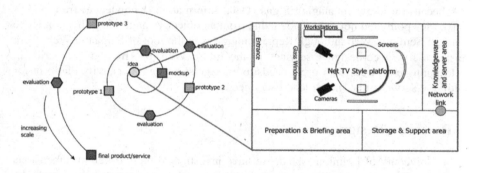

Fig. 7. The creativity or co-design platform to experiment future products/services

From a pragmatic viewpoint (see Fig. 7 on the left), a Creativity Platform is a meeting place and communication space for researchers and developers, businesses and users to pursue innovative and useful projects. For example, they are aimed at collectively defining the characteristics of e-services in order to ensure the most direct correspondence between expectations and use, by combining design and use approaches. Sign sharing makes use of the Creativity Platform by applying an iterative assessment process with end-users from the idea to the product/service through mock-ups and prototypes evaluation.

Co-design Methodology. A Creativity Platform is also called a Co-design Platform [28]. We conceived our own method of co-designing e-services on this Platform (Fig. 8). From ideas to final product/services, resolutions are taken in the frame of funded projects that define the vision, the objectives, the plan and the evaluation phases (production process in a co-working mode on the left of Fig. 8). Assessments of future e-services are made by considering the usage side of the project: end-users have their own identities, activities, tasks and give meaning to the obtained results (usage process in a co-learning mode on the right of Fig. 8). To facilitate the decisions, the project managers (lead-users) are themselves practitioners of the domain. So they have an understanding of the solutions that they can design, deliver and experiment with other users.

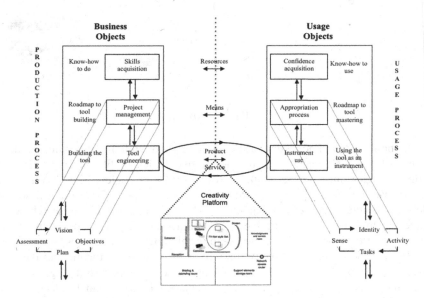

Fig. 8. Business and usage objects of the co-design methodology on a creativity Platform

We will illustrate this co-design methodology used on a Creativity Platform with two examples taken in biodiversity informatics and in music instrumental e-learning.

6 Findings

Fifteen years ago at University of Reunion Island, some researchers of the mathematics and computer science team (LIM) decided to apply knowledge engineering and human-computer interaction to the enhancement of insular tropical environment and instrumental e-learning.

Biodiversity Informatics. For biodiversity *knowledge management* or computer-aided taxonomy, we developed an Iterative Knowledge Base System platform called IKBS [29] with some taxonomists for modeling and describing their collections of corals (Fig. 9). Knowledge bases are the written descriptions of collection specimens that can be compared with dissimilarity measures for classification or identification purpose.

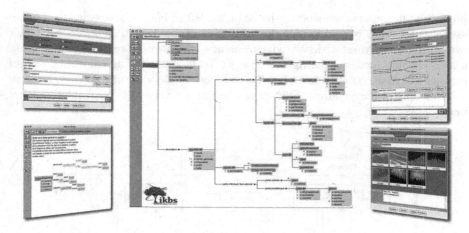

Fig. 9. IKBS

It is based on a knowledge acquisition method and an observing guide for describing biological objects, i.e. the descriptive logics in life Sciences [30]. Our descriptive logics must not be confused with description logics (RDF, OWL) of the *Semantic Web* because they are the rules of thumb of experts for making descriptive models (ontologies) and describing cases. The objective of this Research tool in Biodiversity Informatics is to help biologists classify and identify a specimen correctly from an expert viewpoint by using onto-terminologies (ontologies + thesaurus). With IKBS, experts can use directly the tool as an instrument for making the job of knowledge transmission, i.e. build a knowledge base and illustrate it (Fig. 10).

But the teaching of biological objects appeared to be a real bottleneck in the decision help process. The interpretation problem of specimen descriptions made by specialists emerges when the knowledge base is put in the hands of other biologists: these subjects are not able to observe and describe biological objects with the same know-how and accuracy, thus leading to wrong identifications. On the usage side of co-design, a new co-learning method was then required based on sharing observation know-how rather than transmit knowledge.

Fig. 10. IKBS can be used directly by experts, here on corals

The Learning problem from the end-user viewpoint is to *know how to observe* these objects in order to identify correctly the name of the species. This task is complex and needs help from the specialists who know by experience where to observe correctly the "right characters". By taking care of this knowledge transmission bottleneck, we enter the domain of Sign management for getting more robust results with end-users. Our idea of Sign management is to involve end-users with researchers and entrepreneurs for making them participate to the design of the product/service.

The problem we have to face with when making knowledge bases is that their usefulness depends on the right interpretation of questions that are proposed by the system to obtain a good result. Hence, in order to get correct identifications, it is necessary to acquire qualitative descriptions. But these descriptions rely themselves on the observation guide that is proposed by the descriptive model. Moreover, the definition of this ontology is dependent upon easy visualization of descriptive logics.

At last, the objects that are part of the descriptive model must be explained in a thesaurus for them to be correctly interpreted by targeted end-users. Behind each Object, there is a Subject that models this Object and gives it an interpretation. In life sciences, these objects can be shown to other interpreters and this communication between Subjects is compulsory for sharing interpretations, and not only transmitting knowledge (Fig. 11).

In this new frame, the IKBS project (Iterative Knowledge Base System) will become an ISBS project (Iterative Sign Base System). Its aim is to co-design a Sign Base for Biodiversity management (BSB) rather than a Knowledge Base (KB) with the interactions coming from a community of biologists and amateurs on a Creativity Platform. With ISBS, teachers and learners can play together to share their interpretations of observations. This project will benefit from the long experience we accumulated in the field of Mascarene Corals and Plants identification [31].

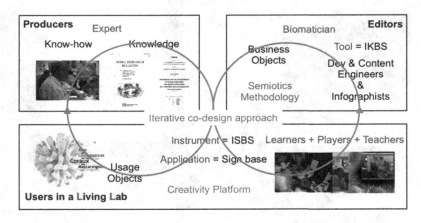

Fig. 11. The sign management process for coral objects' interpretation

The challenge of Sign management for Science observation such as Systematics is to involve all types of end-users in the co-design of Sign bases for them to be really used (e-service). It is why we, as biologists and computer scientist (biomaticians), emphasize the instantiation of a Living Lab in Teaching and Learning at University of Reunion Island for sharing interpretations of objects and specimens on the table rather than concepts and taxa in the head of subjects.

Instrumental e-Learning. Sharing Signs is particularly relevant in artistic fields, where a perfect synchronization between gestures, senses and feelings is essential in order to produce original and beautiful works.

In this frame, the @-MUSE project (@nnotation platform for MUSical Education) aims at constituting a Musical Sign Base (MSB) with the interactions coming from a community of musicians. This project benefits from the experience we accumulated in the field of instrumental e-learning in Reunion Island, from various mock-ups to complete projects such as e-Guitare or e-Piano [32]. Figure 12 sums up our research process in this domain, based on a Creativity Platform [33].

While the different versions of e-Guitare were more centered on the teacher performance, the FIGS (Flash Interactive Guitar Saloon) service was more axed on the dialog between learners and teachers through an online glosses system. What principally emerged from these projects was the need to facilitate the creation and maintenance of new content on the platform. Indeed, while those projects required the intervention of computer scientists and graphic designers in order to create high-quality resources, @-MUSE aims at empowering musicians into creating and sharing their lessons by themselves, on the basis of a common frame of reference: the musical score.

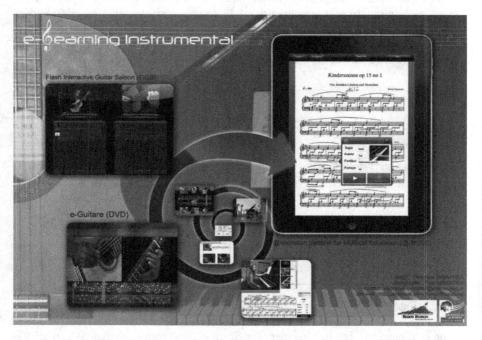

Fig. 12. Instrumental e-learning services co-designed on a creativity platform (CP)

To do so, we designed a MSB. It consists in a set of annotated performances (specimen, or instance) each related to a given musical work (species, or class). This base can be used to compare various performances from music experts or students, and also to dynamically build new music lessons from the available content. To do so, we define a Musical Sign (MS) [34], as an object including a content (a musical performance or demonstration), a form (a score representing the played piece) and a sense (the background experience of the performer, what he or she intends to show) from the viewpoint of a subject (the creator of the Sign).

Figure 13 describes the composition of a MS that can be shared on the platform through a multimedia annotation. Indeed, the principle of @-MUSE is to illustrate abstract scores with indexed multimedia content on top of MusicXML format [35] in order to explicit concretely how to interpret them. Besides, as shown on Fig. 13, multimedia annotations embed all three components of a Sign (data, information and knowledge). This procedure is inspired from a common practice in the music education field, which consists in adding annotations on sheet music in order to remember tips or advice that were validated during the instrumental practice [36].

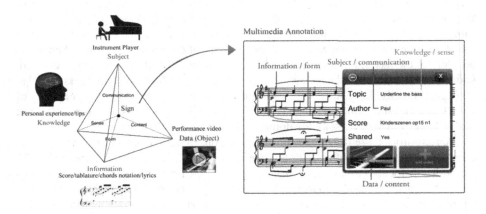

Fig. 13. The musical sign tetrahedron illustrated with a multimedia annotation on @-MUSE

Grounding the @-MUSE service on this practice insures a transparent and natural usage for musicians who already annotate their scores by hand, and additionally enables them to show what they mean using multimedia features. As such, @-MUSE empowers musicians into creating their own interactive scores, using for instance mobile tablets equipped with webcams (@-MUSE prototype [37]).

Collaborative aspects are also essential in music learning, where one progresses by confronting his performances to others'. In this frame, managing Signs rather than Knowledge is particularly relevant, as there is no "absolute truth" in artistic fields: each interpretation can lead to technical discussions between musicians, and their negotiations should be illustrated with live performances to be shown, then understood.

From Knowledge Transmission to Sign Sharing. Knowledge is subjective in the paradigm of Sign management: it cannot be taken for granted without putting it into use, mediated and negotiated with other Subjects on a meeting place, which we called a Creativity Platform. What can be managed is called descriptive or declarative knowledge: it is the communication of justified true beliefs propositions from one Subject made explicit. The formal interpretation process from observation to hypotheses, conjectures and rules is called signification of knowledge on the human communication side of the Sign. It is called representation or codification of knowledge on the machine information side of the Sign. Apart from being described, this interpretation process can be shown with *artifacts* to illustrate the description ("draw me a sheep", says the little prince!). Sign management wants to enhance this aspect of multimedia illustration of interpretations to facilitate transmission and sharing of knowledge through the communication of the Subject (see the fourth communication part of the sign in Fig. 4).

In knowledge management, propositional knowledge is taken mostly in the sense of scientific knowledge, considered as objective in scientific books, and providing the know-that or know-what. Ryle in [38] has shown that this is confusing. In the sense of subjective knowledge taken as "I know that or I know what", there is the other sort of knowledge called know-how. It is "the knowledge of how to do things", i.e. what the subjects can show through their interpretations when they practice their activity (there is a difference between the recipe and the cooking of the recipe, isn't it?). And some people do the activity better than others. They are called the experts. As such, know-how is closer to data (Praxis) and information (*Techne*) than to knowledge (*Scientia*). Finally, know-how and know-that or know-what are different categories of knowledge and should not be conflated [39]. Knowledge synthesizes what makes sense in the head of skilled persons for doing well the tasks of their activity.

Starting from these differences of interpretations about the term of knowledge, and considering the domain of activity that we want to deal with, i.e. education with ICT, we prefer to focus on *managing interpretations*, and firstly the good ones from professors. Sign management manages live knowledge, i.e. subjective objects found in interpretations of real subjects on the scene (live performances) rather than objective entities found in publications (bookish knowledge).

Sign or know-how management produces sign bases that are made of interpretations for knowing how-to-do things with multimedia content and not only knowing what are these things in textual Knowledge bases.

Sign management makes explicit the subjective view of doing arts and sciences. Our aim is to compare different interpretations of subjects about objects through transmitting and sharing them on a physical and virtual space dedicated to a special type of e-service, i.e. in instrumental e-learning or biodiversity informatics.

As shown in music and biodiversity Teaching and Learning, if we want to innovate with people, we should use the concept of Sign management rather than Knowledge management, because the paradigm shift is to pass from knowledge transmission to sign sharing by managing know-how.

Since several years in computer-aided systematics, we proposed a knowledge management methodology based on a top-down transmission of experts' knowledge, i.e. acquisition of a descriptive model and structured cases and then processing of these specimens' descriptions with decision trees and case-based reasoning. We designed a tool called IKBS for Iterative Knowledge Base System to build knowledge bases. But the fact is that Knowledge is transmitted with text, not shared with multimedia, and there is a gap between interpretations of specialists and end-users that prevents these lasts from getting the right identification.

More recently in instrumental e-learning, we focused on the need to show gestural know-how with interactive multimedia contents to play correctly a piece of music, by annotating electronic scores with @-MUSE. This pedagogical approach is based on a gloss system on the Web that can be indexed in codified musical notation.

Today, we prefer to deliver a Sign management method for Teaching and Learning how to identify these collection pieces (specimens or scores) on a Co-Design or Creativity Platform. This bottom-up approach is more pragmatic and user-centered than the previous one because it implicates end-users at will and is open to questions and answers. The role of biological and musical experts is to show amateurs how to play, observe, interpret and describe these art and science works. The responsibility of semioticians (the new cogniticians) is to store and share experts' interpretations of their observation and playing, i.e. know-how rather than knowledge in sign bases with multimedia annotations for helping them to define terms, model their domain, and allow end-users to interpret correctly the objects.

7 Conclusion

As computer scientists and knowledge engineers, we want to design a new Iterative Sign Base System: ISBS = IKBS+@nnotation.

It will be the kernel of our Information Service for defining ontologies and terms, describing pieces work, classifying them with machine learning techniques, and identifying the name through a multimedia interactive questionnaire. The objective of such a tool is to become an instrument in users' hands for monitoring biodiversity in the fields with the National Park of Reunion Island [40], and music at home with the Regional Music Conservatory [41].

For achieving this, we stressed on the importance of reducing the gap between interpretations of teachers (specialists) and learners (amateurs) to get the right identi-fication name and then access to information in databases, or to get the correct gesture that gives the right sound for playing music. This pedagogical effort must concretize itself on a Co-Design or Creativity Platform, which is the Living Lab meeting place for teachers, players and learners, and where these people can manipulate the objects under study, test the proposed e-services and be guided by experts' advices. The teacher is a *producer* who communicates his skilled interpretation of an activity at different levels of perception: psychological motivation, training action, and reasoning feedback. The players are designers-developers *editors* that produce multimedia contents of the expert tasks to perform a good result and index them in a sign base. The learners are *pro-sumers* (producers and consumers) who experiment the sign bases on the physical or virtual Co-Design Platform and tell about their use of the tool to domain experts, ergonomists and anthropologists, in order to improve the content and the functionalities of the mock-ups and prototypes.

Behind each Object to observe, play and describe, there is a Subject who expresses himself and interprets an object by adding his proper signification. This is why we differentiate the Semantic Web, which is the business object approach (the Web of things) represented "objectively" with some description logics (formal syntax for ontologies and cases), and the Semiotic Web that is the usage object approach (the Web of Signs) signified by some descriptive logics of the domain (meaningful process of performance), and which are more subjective. The purpose of the Semiotic Web is to facilitate a consensus between community members, without forgetting that some interpreters are smarter than others in performing a Science or an Art. Their expertise will be visible if users show their interpretations of objects by multimedia artifacts (HD video, 3D simulation, annotated drawings or photos), and if other end-users can ask questions on their know-how and negotiate interpretations. It is why in the frame of natural and cultural heritage enhancement, we proposed to develop Teaching and Learning by Playing e-services with people using Sign management on a Co-design Platform in a Living Lab at the University of Reunion Island [42].

In the post-industrial age of our digital society, designing new services on the Web is crucial for regional territories in order that they become more attractive, competitive, and also more sustainable in the global economy. But up to now, innovation is mainly seen as a linear technological downstream process, centered on enterprises (clusters) and not viewed as an iterative usage upstream process, focused on individuals (Living Labs).

The *form* of LL is attractive because it is an ecosystem based on democratizing innovation with people. User-centered design innovation means that some people, called lead-users, want to innovate for themselves. It has been shown that these persons make most of the design of new services, and only a few come from manufactures.

The *content* of LL is competitive because the best solutions from lead-users are experimented in real time by making situational analyses in "usage laboratories". Mock-ups and prototypes are tested and instrumented to get the best-customized-personalized products and services. For example, the game design (user interaction) and interfaces of 3D multimedia video games benefit greatly from the analysis of feedbacks coming from end-users in communities of practice. So, the success of the e-service does not depend only on the technical success: it has more to do with the quality of human-computer interaction provided with the technology.

At last, the *sense* of LL should be more sustainable, i.e. to render a useful and free service before being profitable, i.e. not only based on a monetary basis but also on trust and reputation. This characteristic is fundamental in the meaning of open access innovation to serve a mission within the scope of products and services made by publicly funded universities. The ultimate value would be to create a form of digital companioning in order to reposition human sharing at the core of technology race.

References

1. http://en.wikipedia.org/wiki/Lisbon_Strategy
2. Kok, W.: Facing the challenge, the Lisbon strategy for growth and employment, Report from the High Level Group (2004)

3. Luyckx Ghisi, M.: The Knowledge Society: A Breakthrough Towards Genuine Sustainability, Editions India, Arunchala press, Cochin, Kerala (2008)
4. Curley, M., Salmelin, B.: Open Innovation 2.0: A New Paradigm, OISPG White Paper (2013)
5. http://www.event.fi-poznan.eu/fia/page/1499/
6. http://www.ictusagelab.fr/ecoleLL/content/literature
7. http://www.youtube.com/watch?v=8HATwHQMb6I
8. Bergvall-Kåreborn, B., Ihlström, E.C., Ståhlbröst, A., Svensson, V.: A milieu for innovation: defining living labs. In: 2nd ISPIM Innovation Symposium: Simulating Recovery - The Role of Innovation Management, 6-9 December, p. 12, New York (2009)
9. Følstad, A.: Living labs for innovation and development of information and communication technology: a literature review. Electron. J. Virtual Organ. Netw. **10**, 99–131 (2008). special issue on Living Labs
10. Santoro, R., Bifulco, V: ESoCE-NET White Paper: The "Concurrent Innovation" paradigm for Integrated Product/Service Development (2006). [http://www.esoce.net]
11. European Commission: Open innovation 2.0 Yearbook 2013, Luxembourg Publications Office of the European Union (2013). [https://ec.europa.eu/digital-agenda/node/66129]
12. http://eit.europa.eu/eit-community/eit-glance/mission
13. http://eit.europa.eu/eit-community/eit-digital
14. European Institute of Innovation & Technology: Catalysing Innovation in the Knowledge Triangle – Practices from the EIT Knowledge and Innovation Communities, Technopolis group (2012)
15. Markkula, M.: The knowledge triangle renewing the university culture. In Lappalainen, P., Markkula, M. (eds). The Knowledge Triangle - Re-Inventing the Future. European Society for Engineering Education, pp. 11–31, Aalto University, Universidad Politecnica de Valencia (2013)
16. Adamson, L., Flodström, A.: Teaching for quality in the knowledge triangle – european institute of innovation and technology's (EIT) coming quality assurance and learning enhancement model. In: Conference proceedings The future of Education, Florence, 16-17 June (2011)
17. Peirce, C.S.: Elements of logic. In: Hartshone, C.H., Weiss, P. (eds.) Collected Papers of C. S. Peirce (1839-1914). The Belknap Press, Harvard Univ. Press, Cambridge (1965)
18. von Uexküll, J.: Theoretical Biology. pp. xvi+362. Kegan Paul, Trench, Trubner & Co., London (1926) (Transl. by D.L. MacKinnon. International Library of Psychology, Philosophy and Scientific Method)
19. Farreny, H.: Les Systèmes Experts: Principles et Exemples. Cepadues-Editions, Toulouse (1985)
20. Shapiro, J.A.: Bacteria are small but not stupid: cognition, natural genetic engineering and sociobacteriology. Stud. Hist. Philos. Biol. Biomed. Sci. **38**(4), 807–819 (2007)
21. Maslow, A.H.: A theory of human motivation. Psychol. Rev. **50**(4), 370–396 (1943)
22. Engeström, Y.: Learning by expanding: an activity-theoretical approach to developmental research. Orienta-Konsultit Oy, Helsinki (1987)
23. Barbieri, M.: Introduction to Biosemiotics: The New Biological Synthesis. Springer, Heidelberg (2007)
24. Sebeok, T.A., Umiker-Sebeok, J. (eds.): Biosemiotics: The Semiotic Web 1991. Mouton de Gruyter, Berlin (1992)
25. de Saussure, F.: Nature of the linguistics sign. In: Bally, C., Sechehaye, A. (eds.) Cours de Linguistique Générale. McGraw Hill Education, New York (1916)

26. University of Reunion Island Living Lab vision (2011). http://www.slideshare.net/conruyt/urlltl, http://www.openlivinglabs.eu/livinglab/university-reunion-island-living-lab-teaching-and-learning
27. New Media Consortium (2013). http://www.nmc.org/publications
28. Sébastien, O., Conruyt, N., Grosser, D.: Defining e-services using a co-design platform: example in the domain of instrumental e-learning. J. Interact. Technol. Smart Educ. 5(3), 144–156 (2008). ISSN 1741-5659
29. Conruyt, N., Grosser, D.: Knowledge management in environmental sciences with IKBS: application to systematics of corals of the mascarene archipelago. In: Brito, P., Cucumel, G., Bertrand, P., de Carvalho, F. (eds.) Selected Contributions in Data Analysis and Classification, pp. 333–343. Springer, Heidelberg (2007). ISBN 978-3-540-73558-8
30. Le Renard, J., Conruyt, N.: On the representation of observational data used for classification and identification of natural objects. In: Diday, D., Lechevallier, Y., Schader, M., Bertrand, P., Burtschy, B. (eds.) New Approaches in Classification and Data Analysis. Studies in Classification, Data Analysis, and Knowledge Organization, pp. 308–315. Springer, Heidelberg (1994). ISBN 978-3-540-58425-4
31. http://coraux.univ-reunion.fr/, http://mahots.univ-reunion.fr/
32. http://e-guitare.univ-reunion.fr/e-piano.univ-reunion.fr/
33. Conruyt, N., Sébastien, O., Sébastien, V., Sébastien, D., Grosser, D., Calderoni, S., Hoarau, D., Sida, P.: From knowledge to sign management on a creativity platform, application to instrumental e-learning. In: 4th IEEE International Conference on Digital Ecosystems and Technologies, DEST, Dubaï, 13-16 April (2010)
34. Sébastien, V., Sébastien, D., Conruyt, N.: Dynamic music lessons on a collaborative score annotation platform. In: The Sixth International Conference on Internet and Web Applications and Services, ICIW, St. Maarten, Netherlands Antilles, pp. 178–183 (2011)
35. Castan, G., Good, M., Roland, P.: Extensible markup language (xml) for music applications: an introduction. In: Hewlett, B.W., Selfridge-Field, E. (eds.) The Virtual Score Representation, Retrieval, Restoration, pp. 95–102. MIT Press, Cambridge (2001)
36. Winget, M.A.: Annotations on musical scores by performing musicians: collaborative models, interactive methods, and music digital library tool development. J. Am. Soc. Inf. Sci. Technol. 59(12), 1878–1897 (2008)
37. Sébastien, V., Sébastien, P., Conruyt, N.: @-MUSE: Sharing musical know-how through mobile devices interfaces. In: 5th Conference on e-Learning Excellence in the Middle East, Dubaï (2012)
38. Ryle, G.: The Concept of Mind. Hutchinson, London (1949)
39. Callaos, N.: The essence of engineering and meta-engineering: a work in progress. In: The 3rd International Multi-Conference on Engineering and Technological Innovation (IMETI 2010), Orlando, 29 June–2 July (2010)
40. http://www.reunion-parcnational.fr
41. http://conservatoire.regionreunion.com
42. www.slideshare.net/conruyt/living-lab-and-digital-cultural-heritage

The Role of Data Warehouse as a Source of Knowledge Acquisition in Decision-Making. An Empirical Study

Moh'd Alsqour and Mieczysław L. Owoc(✉)

Wroclaw University of Economics, Komandorska 118/120, 53-345 Wrocław, Poland
mohsqour@wp.pl, mieczyslaw.owoc@ue.wroc.pl

Abstract. The main purpose of conducting this research is to investigate empirically the role and importance of data warehouse (DW) in enhancing the effectiveness of decision-making. It is believed and assumed that meaningful and significant information can be acquired from DW which provides valuable knowledge to support business process and decision-making. A mail questionnaire survey was regarded as the appropriate method for gathering data. The questionnaire was developed based on the findings from related literature and other related research questionnaires. All the firms (277), which were listed on Amman Stock Exchange (ASE), at the time of data collection, were selected. The researchers arrived at scores of significant and remarkable results regarding DW and its role in enhancing the process of decision-making. The survey's findings showed that the percentage of implementing DW in the Jordanian firms involved is 35 %. In general, the respondents had a positive attitude towards the implementation of DW.

Keywords: Data warehouse · Decision-making · Information · Decision-makers

1 Introduction

Today's organizations face a very hard time, largely as a result of competition, globalization, automation and scarcity of resources. As the business environment is changing. Companies rely more and more on changing technology. At the same time, those companies are likewise evolving. In this changing environment, companies are much more eager in getting immediate and accurate information to make better decisions. Successfully supporting managerial decision-making has become critically dependent upon the availability of integrated and high quality information organized and presented to managers in a timely and easily understood manner [1].

Nowadays there is a need for more advanced and accurate systems. Today's systems are more likely to produce more accurate information than previous systems. Traditional systems, indubitably, are more likely to produce less accurate data and information which lead to bad and incorrect decisions. However, the situation will be different if the firms implement data warehouse (DW) technology, which has emerged as a key source and powerful tool for delivering and accessing information for decision-makers [1–4]. Since the 1990s [5], DWs have been an essential information technology (IT) strategy component for large and medium-sized global organizations [5].

© IFIP International Federation for Information Processing 2015
E. Mercier-Laurent et al. (Eds.): AI4KM 2014, IFIP AICT 469, pp. 21–42, 2015.
DOI: 10.1007/978-3-319-28868-0_2

Large organizations are facing significant challenges in maintaining an integrated view of their business [6]. In an ever more complex and competitive world, the complexity of the organizational context and the management task involving decision-making and assessment of information has increased [7]. Moreover, the changing business environment has an impact on the nature of decisions and decision-making drivers [12]. Timely and informed decision-making is becoming crucial for the long-term success of businesses [13]. [14] claims that business decisions must be made with speed and accuracy if organizations are to remain competitive. Decision-making in those environments involves large data volumes and includes a wide variety of decision tasks [8]. In such environments it is important to assure decision-makers of the quality of data they use [8, 13].

There is quasi-consent that DW provides more detailed and accurate information for decision-makers to improve their decisions. DW systems are perceived to be important tools in the modeling of that complexity. [10] claim that DW hastens the process of retrieving information needed for decision-making. However, reports of high failure of DW systems are common [5, 7, 9].

Despite the recognition of data warehousing as an important area of practice and research, there is little empirical research [1] about implementation of DW in general [1, 9, 11]. Considering the usefulness of DW there has been little research in Jordan on DW, i.e. it has been comparatively less investigated in Jordan. Therefore, the main focus of this study is on the advantages of DW as a provider of information to the process of decision-making. It investigates mainly the relation between decision-making, the need for information and the employment of DW in Jordanian firms. In addition, this study investigates how top management of Jordanian firms perceives the effectiveness (usefulness) of DW as a source of reliable and accurate information for decision-making. Although many studies to DW have been published, they have been concerned with technical issues. However, it has been understood recently that the information systems (IS) failure is due to psychological, environmental, organizational issues etc. rather than technological issues, hence individual differences must be addressed [15]. Overall, there is a scarcity of empirical studies that examine the data warehousing success within an integrative model [9].

2   The Study's Aims

In this research paper, a field study of DW and its role in easing and enhancing the process of strategic decision-making among top managers in Jordanian firms were investigated. Therefore, the main aim of this study is to investigate the role, which is played by DW, in decision-making. In realizing this aim, the researchers believe that the following matters in particular deserve careful investigation largely due to their close connection with the aim and its achievement:

1. To investigate the relationship between the implementation of DW by Jordanian firms and the effectiveness of strategic decision-making (assuming that there is a positive association between the implementation of DW and the degree of decision performance).

It has been ascertained that DW is superior to traditional database and improves the process of decision-making. Therefore, one of the study's aims is to investigate empirically whether or not the DW provides better and more accurate information. It is anticipated in this study that the enhancement of decision-making's process might drive Jordanian firms to employ DW. Thus, another aim of this study is to identify the Jordanian firms' reasons for implementing DW.

2. To identify the grounds of implementing DW by Jordanian firms.

A study of a large number of data warehousing practitioners and experts by [16] showed that the implementation of DW was motivated more by internal pressures than external. A majority of the respondents said that the need was information related. [17] are of the opinion that improving access to information and delivering better and more accurate information are motivations for using DW.

3   The Study's Questions and Hypotheses

The hypotheses are formulated in the light of the study theoretical and conceptual framework and partially based on related previous studies. It is assumed that the outputs of DW, such as data and information, have more positive effect on the process of decision-making by comparison with traditional database systems. This assumption leads the researchers to question whether there is a direct positive relationship between improving the process of decision-making and the implementation of DW? In other words, does the DW provide relevant, reliable and sufficient information for decision-makers to take and achieve sound and effective decisions? This question leads to Hypothesis 1.

Hypothesis 1: There is a strong association between employing DW in Jordanian firms and improving the process of decision-making (soundness and effectiveness of decisions).

In addition to the principal objective of this study, the researchers attempt to achieve multiple aims through this study, including the grounds, which drove Jordanian firms, of implementing DW. This aim led to question what the Jordanian firms' grounds of implementing DW are. This question, in turn, led the researchers to formulate Hypothesis 2.

Hypothesis 2: The dearth of reliable information for taking decisions, the highly competitive environment, the need for more accurate, reliable, relevant and timely information, the changes in manufacturing technology, techniques and processes, the unreliability of existing systems of decision support and inability of the existing systems of decision support to provide reliable, useful and relevant information to the process of decision-making are the major grounds of implementing DW in Jordanian firms.

4   The Importance of the Study

This study was applied in Jordan, which is one of the developing countries in the Middle East. As Jordan's firms are not in isolation from the rest of the world, they are also influenced by the current competitive environment. The implementation of advanced and recent innovations, such as DW, is essential for the firms in developing economies, such as Jordan. Therefore, investigating their implementation in Jordanian firms is well worth considering. However, a few empirical studies, which had been identified on Jordanian firms, prompted the researchers to examine whether this innovation has been successfully implemented in Jordan. Because of the need for further investigation on this topic, considering its importance and the shortage of the empirical research, the researchers have every reason to investigate this issue which has not been given the proper attention in Jordanian environment. Despite the exaltation and adulation of DW, There is a need for evidences that the implementation of DW improves the quality and accessibility to information. Therefore, this study aims to practically investigate whether or not the implementation of DW improves the quality and accessibility to information and enhances the process of decision-making. [18] drew attention to the role and importance of information accessibility. The authors claim that the information accessibility is a precursor of information quality-it has a significant impact on the information's usage, and consequently is an indicator of the DW's success in storing and processing information.

In addition to information's quality, previous literature and studies on DW have emphasized the importance of accessing information. The accessibility to information and their quality are crucial to the success of their use. It has been claimed that the use/ implementation of DW improves the quality and accessibility to information, and consequently leads to sound decisions. In other words, it leads to more fact-based decisions. DW has, according to DW literature, the ability to store a vast amount of data in a usable and appropriate form for the decision-makers' needs and uses. Although a wide range of primary and secondary sources has emphasized the importance and role of information quality and accessibility in enhancing the process of decision-making, little empirical research has been conducted so far. Such claims need to be tested empirically. It is essential, therefore, that the researchers investigate whether or not DW provides easy access to data and information, frequent, accessible and timely reports and more accurate, useful, reliable, complete and relevant information to decision-makers.

Based on extensive literature review, the researcher has identified that firms are often unsuccessful due to a lack of appropriate information or more precisely their inability to get the right information to the right person at the right time. The availability of apposite information to decision-making helps managers in taking reliable decisions, which improve the firm's performance. For this reason, this study is one of the few empirical studies (e.g. [4]), which attempts to examine the effect of DW on decision effectiveness. In addition, previous research has not empirically tested its effectiveness in DSSs contexts in Jordanian firms "to the researchers' knowledge".

Additionally, the researchers have not found and are completely unaware of any empirical studies regarding the implementation of DW in Jordan. Therefore, it is hoped that the findings of this study give the readers and those who are interested in these issues practical insights into DW's field in Jordan. To some extent, it is one of the academic contributions. It contributes to our understanding of the DW in general and in Jordan in particular, and may form a basis and motivation for future research in the important fields. It is also believed that the final outcome of this paper adds up to the improvement and development in DSS, such as DW, by helping their users and developers to be more aware of the data and information's quality.

5 Research Methodology

The study is passed in different phases. In the initial phases of the study, renowned journals, publications, conferences proceedings and books were reviewed. In addition to these sources, the findings of numerous empirical studies were researched and analyzed. Based on the literature review the study' hypotheses were formulated. As the researchers previously pointed out, the sample of the study comprises all the 277 firms, which are listed on ASE at the time of the data collection. Thus, a questionnaire was found to be the best instrument for collecting the data in this study. During the next phases, therefore, a survey questionnaire was conducted with the top managers of Jordanian firms. The questionnaire, which is used in this study, is based on previous studies and the researchers' assessments and discretion and adapted to suit the objectives and requirements of the study. The primary objective of the questionnaire is to amass appropriate data and responses from the potential respondents for testing the significance of hypotheses.

In order to answer the study's questions, fulfill its aims and find out whether or not there is a positive relationship between independent and dependent variables, the study's hypotheses were statistically tested. The data, which were collected, is very quantitative in nature.

Therefore, in the final phases of the study, the data were statistically analyzed by employing Statistical Package for the Social Sciences (SPSS) in order that proper descriptive and inferential statistics to analyze the results and draw conclusions can be reached, including means, frequencies, standard deviation, t-test, f-test and chi-square.

6 Literature Review

The concept of data warehousing has evolved out of the need for easy access to a structured store of quality data that can be used for decision-making [19, p. 5]. Organizations have vast amounts of data but have found it increasingly difficult to access it and make use of it. This is because it is in many different formats, exists on many different platforms, and resides in many different file and database structures, and, as a result, organizations have had to write and maintain perhaps hundreds of programs that are used to extract, prepare, and consolidate data for use by many different applications for analysis and reporting [19, p. 5].

As an attempt to solve the problem, DWs were introduced. DWs have become the focal point for decision support in organizations today [20] and emerged as a key platform for the integrated management of decision support data in organizations [3].

[19, p. 5] claim that the data warehousing offers a better approach. Data warehousing implements the process to access heterogeneous data sources; clean, filter, and transform the data; and store the data in a structure that is easy to access, understand, and use. The data is then used for query, reporting, and data analysis [19, p. 5]. [21] also claim that the data warehousing has emerged as an effective mechanism for converting data into useful information.

DW systems offer efficient access to integrated and historical data from heterogeneous sources to support managers in their planning and decision-making [22]. [23] also claim that data warehousing provides an infrastructure that enables businesses to extract, cleanse, and store vast amounts of data. According to [19, p. 1], businesses of all sizes and in different industries, as well as government agencies can realize significant benefits by implementing a DW. Most medium to large organizations, according to [24], operate DWs.

It has been claimed that the main DW's role is to support decision-making. However, the role of DWs has been broadened. [25] state that DW provides information from external data sources for decision-making. DW has the potential to create radical changes to existing business processes and is often viewed within the context of business process reengineering [11]. Accordingly, [26] claims that DW gives business' users the ability to analyze data. [27] also claim that DWs enable organizations to exploit decision-making.

According to [28], DWs provide the basis for management reports and decision support. They added that the purpose of DWs is to take that vast amount of data from many internal and external sources and present them in meaningful formats for making better decisions. In support of the above mentioned claims participants to a study by [3] agreed that the Return on Investment (ROI) for the DW was well justified through considerable gains in productivity and enhanced quality of customer service. Moreover, in an independent detailed study of 62 organizations worldwide [34], the major findings of International Data Corporation (IDC) based upon 62 case studies of organizations that have successful DWs in use are an average three-year ROI of 401 % was realized by organizations building DWs. Although this study is primarily focused on quantitative information, there are several qualitative benefits [34], such as providing standardized, clean and value-added data to create information from disparate sources. In addition, the DW makes the data available across corporate organizations and provides the needed information quickly.

The DW is developed in order to support the integration of external data sources [29] for the purpose of advanced data analysis. [40, p. 35] argues that a DW produces tangible impacts to the quality of day-to-day business transactions. Previous research on DW has produced some encouraging findings about its benefits and indicated that a DW can offer several benefits to an organization [11], such as enabling effective decision support; ensuring data integrity, accuracy, security, and availability; easing the setting and enforcing of standards, facilitating data sharing, and improving customer service [35]. [30] presented time savings for data suppliers and for users, more and better information, better decisions, improvement of business processes, and support for the accomplishment of strategic business objectives as benefits from data warehousing. Furthermore, [2], who examined data warehousing at the Housing and Development Board (HDB) in Singapore, found that the main benefits of the DW, which were developed by HDB, are enabling the users to have access to consistent and reliable data in a timely fashion which facilitated forecasting and planning efforts and improved decision-making. In addition, a study by [31] revealed that DW appears to be used more to improve the flow of information in an organization than to change the way the organization does business. The authors found that more and better data is the greatest realized benefit from DW. Moreover, a study by [32] identified time savings, new and better information, and improved decision-making as benefits of DW.

[33] conducted an explanatory case study at a financial services organization to investigate how DW provides decision support to individual decision-makers. The results showed that the organizations successfully automated the retrieval and input of data for front-end users. [40, pp. 33–34], who interviewed people from seven companies, found that the benefits of implementing DW were improving asset management, reducing customer support costs, auditing billing practices, terminating unprofitable product, reducing staff requirements and running the business. [28], who described the DW implementation at Blue Cross and Blue Shield of North Carolina (BCBSNC), claim that the DW had resulted in many organizational benefits, including better data analysis and time savings for users. [34], who looked at the DW of Egypt's Cabinet Information and Decision Support Center, found that the DW provides a lot of benefits to the users, including ease of access to the information, fast and more consistent reports, support the decision-makers and integrating the data from various sources.

Additionally, [4], who conducted a laboratory experiment in 2006, found that the implementation and use of DW improves the DSS users' decision performance, by which he means improving the quality of the DSS by adding a DW can improve information availability and quality and enhance DSS users' decision performance. In conclusion, the study showed that DW can have a positive impact on decision-making. [35], who described an example of implementing DW in medical institutions, found that the DWs provide the users access to important information. [36], who conducted a survey to find out how DW assists decision-making process in healthcare, found that the DW provides better accessibility to data, integrated disparate data sources and improved decision-making. [37] found that all companies, which are studied, recognized some benefits such as cost reduction, reach-out to other markets, increase in sales, time saving in amount and preparation of reports and more effective decision-making based on the obtained information. [38], who conducted two case studies on American Airlines and Hallmark Cards, found the easy to use, speedy information retrieval, more information, better quality information, improved productivity, and better decisions as benefits of DW. [39], who examined the implementation of DW in public security, found that the DW is very important in improving the comprehensive ability of leadership and decision-making. In addition, it quickly and efficiently integrates heterogeneous data sources.

Previous literature on DW, such as [22, 31], claims that the DW does not create value by itself; the value comes from the use of the data in the DW. [22] claim that improved decision-making results from the better information available in the DW. By making the right information available at the right time to the right decision-makers in the right manner, DWs empower the users with the ability to make the right decisions [40]. [31] also claim that this use can result in numerous benefits, including more and better information, improved user ability to produce information and reduced effort by developers to produce information. [41] also maintain that DWs have tremendous potential to present information. The greatest potential benefits of the DW occur when it is used to redesign business processes and to support strategic business objectives [30]. [42] also identified many different measures of success; these include benefits such as data accuracy, useful information, accurate information, ease of use, user satisfaction, time to make decision and increased revenue. However, [43] claims that despite clear evidences that many DW projects have resulted in interesting business benefits, there are also many examples of cost and schedule overruns and dissatisfaction regarding the results from these projects. [44] argue that DW is one of the key developments in the IS field and has plentiful benefits. In addition, [45] indicates that the introduction of a new IS into an organization should deliver multiple benefits.

Since the early 1990s, DWs have become the technology of choice for building data management infrastructures [5] and been investigated and implemented around the world in many areas and by many researchers, authors, and scholars [46]. According to [47, p. 13], the early successful implementation of DW dates back to mid-1980s at ABN AMRO Bank (Netherlands). The author claims that the end-user's needs were the key feature behind the implementation. As a result, those requirements were modeled rather broadly, and all available data was stored in the DW. In fact, in the first few years of general use, its usage had grown at an annual rate of 50 %, and by 1995 the DW had supported some 3,000 end-users. [47, p. 17] also mentioned that a study of 62 DW projects, which was conducted in 1996, showed an average return on investment (ROI) of 321 % for these enterprise-wide implementation in an average payback period of 2.73 years.

In addition, [48], who investigated whether lodging companies are involved with DW technology through a sample of twelve large lodging corporations, found that the most of hotel corporations in the study were using their DWs to support market analysis. However, [13], who conducted a survey on a large Australian public organization, found that 60 % of the users were with limited or no usage (or were anticipating the use in the future) of the DW. The data also helped the users to make informed decisions and the data, which was retrieved from the DW, was also presented to the senior management and other strategically oriented sections in the form of reports i.e. annual and quarterly reports.

Similarly, [11], who surveyed DW's managers and data suppliers from 111 organizations in different regions of the United States (US), also found that all companies had operational DW and nearly all of them considered that their initiative is successful. In other words, 26 % of the respondents considered it runaway success. Another survey by Forrester showed that 878 IT decision-makers in the US enterprises were somewhat satisfied with the accessibility and quality of customer information; 82 % of these respondents are satisfied or very satisfied [49]. [27] also found that about 55 % of 196 respondents firms (107) in two major states in the US had already adopted and used DWs. Similarly, [23, pp. 192–196) found that 60 % of the respondents consider the functionality of their DW below expectations. 40 % of the dissatisfied group was actually still using it. This means that 80 % of the respondents were still using their internal DW, which is a good indication of the overall degree of satisfaction.

In addition to those studies, [50], who reported the results of a survey which was conducted at The Data Warehousing Institute (TDWI) World Conference in New Orleans 2003, found that 45 % of the respondents had been already in production with their current DW or implementing a second or third release. In Taiwan, [51], who conducted a survey on Taiwanese banks, found that 53.33 % of 30 respondent banks had DW functions or capabilities, 31.25 % (5) out of 16 respondent banks had already implemented the DW and 68.75 % (11) were still in the process of development to implement DW. 85.71 % (12) of 14 banks which had not implemented the DW were evaluating the possibility and/or potential of adopting DW. Another study in Asian countries was conducted by [52]. [52], who surveyed 115 users of DW in four Korean financial companies, also found that the participated companies had been using their data warehousing systems for approximately two years. In addition, the findings also showed that all the respondents were end-users of DWs in their companies, using the systems mainly for financial analyses. The majority of data warehousing users (approximately 70–75 %) use the DWs regularly, i.e. every day.

The 2007's IBM Data Warehousing Satisfaction Survey showed that 56 % of those questioned were very successful with their DW (200 end-user enterprises were participated), however, 43 % of the respondents acknowledged the need for improvement, as there are still a number of business and technical challenges confronting the enterprises which make use of DWs according to the survey [53]. The success of DW's implementation, according to [53], is growing as 56 % of the respondents were very successful and satisfied. [54], who conducted a survey on 84 users of DW, found that the majority of respondents (73 %) in the surveyed firm were successful in obtaining and accessing the needed data and information from the DW, only two respondents indicated that they were not at all successful, while 56 % indicated somewhat successful and 13 % indicated very successful. In addition to these results, 67 % and 33 % of the respondents rated the importance of the obtained information in performing their job better as vital and somewhat important respectively.

Numerous studies have also shown successful implementation of DW. For example, [55] used a case study and conducted a series of interviews at Continental Airlines. The results showed that the organization has realized an enviable level of DW maturity and significant cumulative benefits. [56], who surveyed 244 members of TDWI, found that 51.2 % of the respondents had at least one DW application and 30.3 % were still in development stage. 17.2 % were still in planning stage and 1.2 % of respondents had no any efforts made in their organizations to implement the DW. This result showed that more than 80 % of respondents had implemented DW.

According to [9], there is a scarcity of empirical studies that examine the DW success. In the study population, i.e. Jordan, there is not a shred of empirical evidence that the DW has been investigated or showed the degree of DW implementation, to the best of the researchers' knowledge. However, the literature review was limited to materials which have been published in English language only–evidence in languages other than English could be possible. Therefore, this paper investigates the extent to which Jordanian firms implemented DW and the reasons for implementing DW. It is aimed at providing empirical evidence, thereby extending the body of research regarding the implementation of DSSs in general and DW in particular.

7   Findings and Discussion

As mentioned earlier, the study's sample consists of the Jordanian firms' top management. [57, p. 142] claim that although the large populations are referred to as ideal populations, sampling every person in these populations is not realistic or doable. They believe that time, money, and other restrictions make it impossible for the average survey researcher to reach all members of an ideal population. Therefore, the researchers have to forgo these grand expectations and select a smaller or realistic population. This argument forms the basis for this study's population, therefore, the study's population, which is selected, is only the Jordanian firms that are listed on ASE.

It is believed that this category is more likely to use the outputs of DW for strategic decision-making, i.e. the data, which is obtained from these respondents, fulfill the purpose of this study. Therefore, the potential respondents were solicited for their opinions. The covering letter was addressed to the top management of each firm. The questionnaire, along with a self-addressed stamped envelope (SASE) and a covering letter were posted to the potential respondents. The potential respondents were requested to return the completed questionnaires in the enclosed SASE.

As the main aim of this study is to investigate the role of DW in strategic decision-making, the choice of the target population (Jordanian firms which are listed on ASE) and potential respondents (top management) is made on the basis that those firms have sufficient resources to implement such expensive and time-consuming systems and the respondents have sufficient knowledge about their firms and take the strategic decisions. In other words, they are the main consumers of the DW reports. In addition, the companies' Guide by ASE is the only listing that specifically covers all sectors and industries in Jordan. This directory lists the names, titles, and the general information about the listed companies. Top management of Jordanian firms, which are listed on ASE, constitutes the population of this study. A total of 277 questionnaires were sent out by post along with a covering letter. The potential respondents were assured that all data, which would be provided by them, would be treated by complete confidentiality.

The results of the study's sample analysis are shown in the figure below (Fig. 1). In this figure, the sample results are broken down by responses. As can be seen from these results, 140 completed questionnaires were returned to the researcher's address with a response rate of 50.5 %. According to these figures, usable questionnaires accounts for 43.3 % of the total sample. 20 of the questionnaires were discarded as unreliable, i.e. there were many essential questions missing from the questionnaires.

To sum up, all the firms (277), which were listed on ASE at the time of data collection, were selected. 140 filled questionnaires were returned generating 50.5 %. This response rate somehow on average comparison to many similar studies such as [58, 59]. [58], who studied 350 Jordanian companies, had a response rate of 30 %. [59], who used a self-administered questionnaire and targeted similar population in Jordan, had a response rate less than 50 %.

The reliability, internal consistency and validity of the Likert scale questions are assessed by using Cronbach's alpha. Figure 2 shows Cronbach's alpha for the two Likert-scale questions. In addition, the table demonstrates the mean, SD, sum of item variances (V) and standard error (SE) for the rating scale questions. It was found that

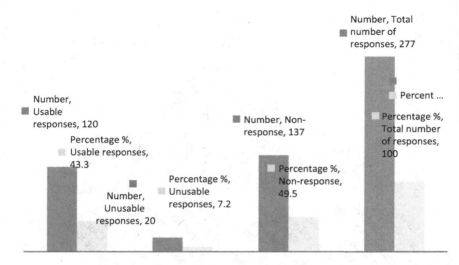

Fig. 1. The analysis of the study's sample (Source: the figures are based on the responses to the survey).

Cronbach's alpha is the most popular method for assessing the reliability of scales. It has been used by many researchers, including [60–71]. Cronbach's alpha determines the internal consistency of the items in a survey instrument (questionnaire) to assess its reliability [72–74].

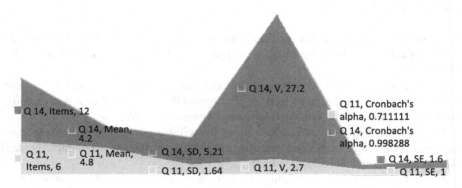

Fig. 2. The analysis of the study's sample (Source: the figures are based on the responses to question 11 and question 12 (Appendix A)).

The types of the firms involved in this survey are illustrated by a bar chart (Fig. 3). In this bar chart, the survey results are broken down by industry group.

As can be seen from these results, the financial services industry is the highest among the others (14.2 %). According to these figures, the insurance industry accounts for 11.7 % of the industry groups. The survey's figures also show that 9.2 % of the firms are within the banking industry, considerably higher than the 0.8 % of firms within the glass and ceramic, textile, leather and clothing and utilities and energy industries and 2.5 %

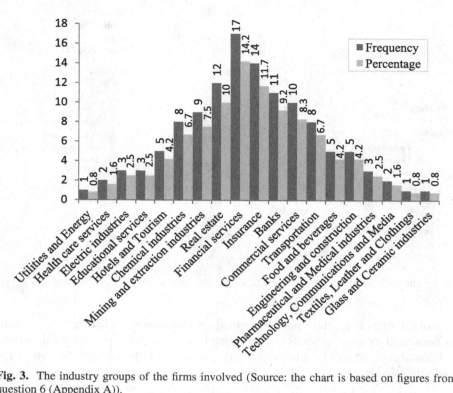

Fig. 3. The industry groups of the firms involved (Source: the chart is based on figures from question 6 (Appendix A)).

within the pharmaceutical and medical, electric and educational services industries. From the data in the above bar chart, it is apparent that almost 9 industries are in the range of 3 to 10 %.

For the purpose of this study, there are only two groups regarding the implementation of DW: first, the firms that have fully implemented DW and used it as a part of daily practices; and, second, the firms that have not implemented DW.

The pie chart (Fig. 4) illustrates how many firms implemented DW. As can be seen from these results, the number and percentage of the firms, which have not implemented DW, are far more than those which implement.

According to this bar chart, only approximately a third (42) of Jordanian firms involved has implemented DW. In other words, 35 % of the firms involved have implemented DW. The firms, which have not implemented DW, make up about 65 % of the 120 firms involved in the survey. The results of the study revealed that the rate of implementing DW in Jordanian firms is very low and less than other countries. For example, [27] found that about 55 % of 196 respondents firms (107) in two major states in the continental US had already adopted and used DWs. [56], who surveyed 244 members of TDWI, found that 51.2 % of the respondents had at least one DW application. However, the results of this study are similar to some previous studies' results. For example, [51], who conducted a survey on 16 Taiwanese banks, found the rate of implementing DW among those banks is 31.25 %.

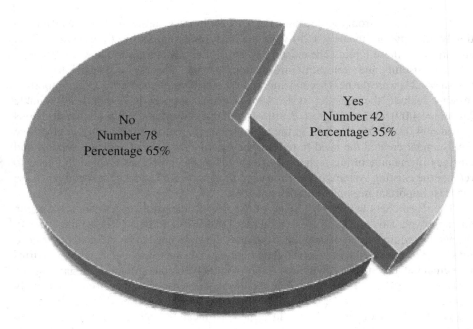

Fig. 4. The percentage of implementing DW among the firms involved (Source: the figures are based on the responses to question 10 (Appendix A)).

The 42 respondents, who their firms implemented DW, were solicited for their opinions regarding the reasons behind the plan (decision) to implement a DW. The likely and expected reasons, which were measured on a five-point scale ranging from 1 (unimportant) to 5 (extremely important), are shown in Table 1. The table shows the mean and standard deviation (SD) for all the select reasons.

Table 1. The Reasons for Implementing DW (Source: the figures are based on the responses to question 11 Appendix A

The reasons for implementing DW	No.	Min.	Max.	Mean	SD
There is a dearth of reliable information for taking decisions	42	2	5	4.07	0.97
The highly competitive environment created the need to replace the existing systems of decision support	42	1	4	2.59	1.11
The need for more accurate, reliable, relevant and timely information	42	3	5	4.28	0.81
Changes in manufacturing technology, techniques and processes created the need to replace the existing systems of decision support	42	1	5	2.47	1.19
The existing systems of decision support have not been reliable	42	2	5	4.09	1.01
The existing systems of decision support have not provided reliable, useful and relevant information to the process of decision-making	42	2	5	4.24	0.93

As can be seen from these results, the need for more accurate, reliable, relevant and timely information (mean = 4.28) were the most important reason to implement DW. According to these figures, the existing systems of decision support have not provided reliable, useful and relevant information to the process of decision-making (mean = 4.24) were the next second important reason to implement DW. Other important reasons include the existing systems of decision support have not been reliable (mean = 4.09) and there is a dearth of reliable information for taking decisions (mean = 4.07). From the data in the table, it is apparent that the highly competitive environment created the need to replace the existing systems of decision support and changes in manufacturing technology, techniques and processes created the need to replace the existing systems of decision support mean = 2.59 and 2.47 respectively were the least important reasons to implement DW.

The researchers believe that some of the keys to measure success are satisfaction and approval. This is why it is essential that DW (if it is to be a success) must satisfy the needs (requirements) of its users. Similarly, the users need to be satisfied of the need for DW. To measure the success of implementing DW, the respondents, who their firms implemented DW (42 respondents), were solicited for their opinions regarding the derived satisfactions from implementing DW. A seven-point scale from 1 (strongly disagree) to 7 (strongly agree) was used to measure these responses. Table 2 shows that the great majority of respondents, since the least mean is 5.43, are very satisfied with the derived benefits of implementing DW. As can be seen from these results, most of the respondents believe that DW is useful in making better business decisions (mean = 6.69).

Table 2. The degree of satisfaction from implementing DW (Source: the figures are based on the responses to question 12 Appendix A

Satisfaction and approval of implementing DW	No.	Min.	Max.	Mean	SD
The information in your firm's DW is sufficient for taking sound decisions	42	4	7	6.14	0.89
You are content with the benefits of your firm's DW	42	4	7	6.26	0.85
The DW's information meets the requirements of your task	42	5	7	6.17	0.86
DW is a reliable source of information	42	4	7	6.21	0.95
DW's information is beneficial to different areas of decisions	42	4	7	6.19	0.83
DW's information is used in different areas of decisions	42	5	7	6.14	0.81
The DW offers user-friendly query capability to decision-makers	42	3	7	5.83	1.10
DW is a reliable information system	42	4	7	6.02	1.04
DW is a user-friendly information system	42	3	7	5.43	1.32
DW is useful in making better business decisions	42	5	7	6.69	0.66
DW is useful for decision maker's task	42	5	7	6.62	0.62
DW can output information much more quickly	42	5	7	6.33	0.72
The benefits of implementing DW exceed its cost	42	4	7	6.14	0.89

As can be seen from these results, the need for more accurate, reliable, relevant and Moreover, the figures show clearly that the involved respondents were very satisfied on the grounds that DW is useful for decision maker's task (mean = 6.62), DW can output information much more quickly (mean = 6.33) and if the users content with the benefits of their firm's DW (mean = 6.26). Other high responses include DW's information met the requirements of the respondents' task (mean = 6.17), DW's information is beneficial to different areas of decisions (mean = 6.19), DW is a reliable source of information (mean = 6.21), the benefits of implementing DW exceed its cost (mean = 6.14) and the DW's information is sufficient for taking sound decisions (mean = 6.14). According to these figures, DW is a user-friendly information system (mean = 5.43) and The DW offers user-friendly query capability to decision-makers (mean = 5.83) were the least responses.

Based on these figures, it can be concluded that the implementation of DW (in Jordanian firms) is very useful and have a great positive effect on the process of decision-making. These results are similar to some previous studies' results on DW, for example, [11, 13].

8 Conclusion

The evidences, which are obtained by analyzing the data from the questionnaires, reveal exceptionally remarkable facts, first of all and to some extent, the percentage of implementing DW is high (35 %) by comparison with other countries. One consequence of implementing DW was the great role of DW's information in enhancing and facilitating the process of decision-making. The results showed that the Jordanian firms benefited greatly from implementing DW. The results also revealed that the DW is a fruitful source of information. Moreover, the implementation of DW proved to be a success through helping decision-makers in taking fact-based decisions. Furthermore, the DW was lauded by the users for the successful use of its information as a basis for decision-making.

This study has humbly contributed to the field of scientific research in general and the field of decision support systems (DSS) in particular in many ways, first of all, the studies on the implementation of DW were nearly all in developed countries. This study was applied in Jordan, which is one of the developing countries in the Middle East. Therefore, the results of this study made a humble contribution to the existing knowledge in the field of implementing DW worldwide in general and in Jordan in particular. Second, there is a need for evidences that the DW improves the quality and accessibility to information. Therefore, this study practically investigated whether or not the implementation of DW improves the quality and accessibility to information and facilitates the decision-makers' tasks. Lastly, based on extensive literature review, the researchers have identified that firms are often unsuccessful due to a lack of appropriate information. For this reason, this study is one of the few empirical studies which have attempted to examine the effect of DW on decision effectiveness. In addition, previous research has not empirically tested its effectiveness in DSS contexts in Jordanian firms "to the researchers' knowledge".

Despite the usefulness and positive contributions of the study's results, these results should be treated and interpreted with caution. In fact, the study's sample included only the Jordanian firms which are listed on ASE. As a consequence, this might severely restrict the generalization of the results. It is believed that the results of this study might have been dissimilar, if all Jordanian firms have been surveyed. Therefore, prospective researchers are recommended to broaden the scope of their investigation to include all Jordanian firms.

Appendix A

Part 1: The personal details and demographics of respondents
1. Please give your name, title and Email: ....
....
Note: This question is optional
2. Please indicate which sex you are.
A. Male
B. Female
3. Which age bracket are you?
A. Less than 40
B. 40 to 45
C. 46 to 50
D. 51 to 55
E. 56 to 60
F. More than 60
4. Please indicate what level of education you achieved.
A. General certificate of secondary education
B. A graduate degree (bachelor's)
C. A postgraduate degree (master's)
D. A PhD degree (doctorate)
5. Please indicate your years of experience at the senior management level.
A. Less than 3 years
B. 3-6 years
C. 7-10 years
D. 11-14 years
E. More than 14 years
Part 2: The company's details
6. Your company belongs to which of the following 22 industrial groups. Please tick the appropriate group.
1. Banks
2. Chemical industries
3. Commercial services
4. Educational services
5. Electric industries
6. Engineering and construction
7. Financial services
8. Food and beverages
9. Glass and ceramic industries
10. Health care services
11. Hotels and Truism
12. Insurance
13. Mining and extraction industries
14. Paper and cardboard industries
15. Pharmaceutical and medical industries

16. Printing and packaging
17. Real estate
18. Technology, communications and media
19. Textiles, leather and clothing
20. Tobacco and cigarette
21. Transportation
22. Utilities and energy
7. Please indicate the number of employees in your entire firm, including all of its branches, divisions and subsidiaries.
A. Less than 100
B. 100-500
C. 501-900
D. 901-1300
E. 1301–1700
F. 1701-2100
G. 2101-2500
H. More than 2500
8. The paid-up capital of your firm is:
A. Less than $1million
B. $1m to less than $10 million
C. $10 m to less than 20 million
D. $20 m to less than $100 million
E. $100m to less than $500 million
F. $500m to less than $1 billion
G. 1bn to less than $10 billion
H. 10 billion or more
9. Please indicate the annual revenue of your firm in the US dollars?
A. Less than $500,000
B. $500,000 to less than $1 million
C. $1m to less than $20 million
D. $20 m to less than $100 million
E. $100m to less than $500 million
F. $500m to less than $1 billion
G. $1 billion or more
Part 3: Questions regarding the implementation of data warehouse (DW)
10. Has your firm fully implemented DW and using it as a part of daily practices?
A. Yes
B. No (please go to question11)
11. How important these reasons in the decision to implement DW? Please check the appropriate box or boxes.

		Unimportant	Slightly important	Averagely important	Very important	Extremely important
A.	There is a dearth of reliable information for taking decisions.					
B.	The highly competitive environment created the need to replace the existing systems of decision support.					
C.	The need for more accurate, reliable, relevant and timely information.					
D.	Changes in manufacturing technology, techniques and processes created the need to replace the existing systems of decision support.					

E.	The existing systems of decision support have not been reliable.						
F.	The existing systems of decision support have not provided reliable, useful and relevant information to the process of decision-making.						

12. To what extent do you agree or disagree with these issues? Please tick the appropriate box or boxes.

		Strongly disagree	Somewhat disagree	Disagree	Neutral	Agree	Somewhat agree	Strongly agree
A.	The information in your firm's DW is sufficient for taking sound decisions							
B.	You are content with the benefits of your firm's DW							
C.	The DW's information meets the requirements of your task							
D.	DW is a reliable source of information							
E.	DW's information is beneficial to different areas of decisions							
F.	DW's information is used in different areas of decisions							
G.	The DW offers user-friendly query capability to decision-makers							
H.	DW is a reliable information system							
I.	DW is a user-friendly information system							
J.	DW is useful in making better business decisions							
K.	DW is useful for decision maker's task							
L.	DW can output information much more quickly							
M.	The benefits of implementing DW exceed its cost							

References

1. Hayen, R.L., Rutashobya, C.D., Vetter, D.E.: An investigation of the factors affecting data warehousing success. Issues Inf. Syst. **8**(2), 547–553 (2007)
2. Ang, J., Teo, T.S.H.: Management issues in data warehousing: insights from the Housing and Development Board. Decis. Support Syst. **29**, 11–20 (2000)
3. Shin, B.: An exploratory investigation of system success factors in data warehousing. J. Assoc. Inf. Syst. **4**, 141–170 (2003)
4. Park, Y.-T.: An empirical investigation of the effects of data warehousing on decision performance. Inf. Manag. **43**, 51–61 (2006)
5. Mukherjee, D., D'Souza, D.: Think phased implementation for successful data warehousing. Inf. Syst. Manag. **20**, 82–90 (2003)
6. Eckerson, W.: Evolution of data warehousing: the trend toward analytical applications. J. Data Warehouse **25**(1), 1–8 (2003)
7. Briggs, D.: A critical review of literature on data warehouse systems success/failure. J. Data Warehouse. **49**(3), 1–20 (2002)
8. Shankaranarayanan, G., Cai, Y.: Supporting data quality management in decision-making. Decis. Support Syst. **42**, 302–317 (2006)

9. Rudra, A., Yeo, E.: Issues in user perceptions of data quality and satisfaction in using a data warehouse - an Australian experience. In: Proceedings of the 33rd Hawaii International Conference on System Sciences, pp. 1–7. IEEE (2000)
10. Hegazy, F., Ghorab, K.: The impact of system support on adoption & diffusion of data warehousing success (2003). http://www.hicbusiness.org/biz2003proceedings. Accessed 31 August 2011
11. Aljanabi, A., Alhamami, A., Alhadidi, B.: Query dispatching tool supporting fast access to data warehouse. Int. Arab J. Inf. Technol. **10**(3), 269–275 (2013)
12. Wixom, B.H., Watson, H.J.: An empirical investigation of the factors affecting data warehousing success. MIS Q. **25**(1), 17–41 (2001)
13. Gimzauskiene, E., Valanciene, L.: Efficiency of performance measurement system: the perspective of decision making. Econ. Manag. **15**, 917–923 (2010)
14. Mansouri, S., Gallear, A.D., Askariazad, M.H.: Decision support for build-to-order supply chain management through multiobjective optimization. Int. J. Prod. Econ. **135**, 24–36 (2012)
15. McKenna, J.P.: Moving toward real-time data warehousing. Bus. Intell. J. **16**(3), 14–19 (2011)
16. Au, N., Ngai, E.W.T., Cheng, T.C.E.: Extending the understanding of end user information systems satisfaction formation: an equitable needs fulfillment model approach. MIS Q. **32**(1), 43–66 (2008)
17. Rasmussen, N., Goldy, P.S., Solli, P.O.: Financial Business Intelligence Trends, Technology, Software Selection, and Implementation. Wiley, New York (2002)
18. Watson, H.J., Haley, B.J.: Data warehousing: a framework and survey of current practices. J. Data Warehouse **2**(1), 10–17 (1997)
19. Gatziu, S., Vavouras, A.: Data Warehousing: Concepts and Mechanisms, Informatik, Informatique **1** (1999)
20. Doherty, N.F., Doig, G.: The role of enhanced information accessibility in realizing the benefits from data warehousing investments. J. Organ. Transform. Soc. Change **8**(2), 163–182 (2011)
21. Ballard, C., Herreman, D., Schau, D., Bell, R., Kim, E., Valencic, A.: Data Modeling Techniques for Data Warehousing, 1st edn. International Business Machines Corporation (IBM Corp) (1998)
22. Watson, H., Fuller, J.C., Ariyachandra, T.: Data warehouse governance: best practices at blue cross and blue shield of North Carolina. Decis. Support Syst. archive **38**(3), 435–450 (2004)
23. Ahmad, I., Azhar, S.: Data warehousing in construction: from conception to application. In: Proceedings of the First International Conference on Construction in the Twenty First Century, Miami, Florida, USA, April 2002
24. List, B., Bruckner, R.M., Machaczek, K., Schiefer, J.: A comparison of data warehouse development methodologies case study of the process warehouse. In: Hameurlain, A., Cicchetti, R., Traunmüller, R. (eds.) DEXA 2002. LNCS, vol. 2453, pp. 203–215. Springer, Heidelberg (2002)
25. Nemati, H.R., Steiger, D.M., Iyer, L.S., Herschel, R.T.: Knowledge warehouse: an architectural integration of knowledge management, decision support, artificial intelligence and data warehousing. Decis. Support Syst. **33**(2), 143–161 (2002)
26. Mannino, M.V., Hong, S.N., Choi, I.J.: Efficiency evaluation of data warehouse operations. Decis. Support Syst. **44**(4), 883–898 (2008)
27. Bębel, B., Eder, J., Koncilia, C., Morzy, T., Wrembel, R.: Creation and management of versions in multiversion data warehouse. In: Proceedings of the 2004 ACM Symposium on Applied Computing, SAC 2004, 14–17 March, Nicosia, Cyprus, pp. 717–723 (2004)
28. Brown, T.: Data Warehouse Implementation with the SAS System, SAS Institute Inc., Dallas (1996). http://www2.sas.com/proceedings/sugi22/dataware/paper132.pdf

29. Ramamurthy, K.R., Sen, A., Sinha, A.P.: An empirical investigation of the key determinants of data warehouse adoption. Decis. Support Syst. **44**, 817–841 (2008)
30. Nilakanta, S., Scheibe, K., Rai, A.: Dimensional issues in agricultural data warehouse designs. Comput. Electron. Agric. **60**, 263–278 (2008)
31. Chmiel, J., Morzy, T., Wrembel, R.: Multiversion join index for multiversion data warehouse. Inf. Softw. Technol. archive **51**(1), 98–108 (2009)
32. Hackathorn, R.: Current Practices in Active Data Warehousing. Bolder Technology Inc. (2002)
33. Watson, H., Haley, B.: Managerial considerations. Commun. ACM **41**(9), 32–37 (1998)
34. Watson, H.J., Goodhue, D., Wixom, B.H.: The benefits of data warehousing: why some organizations realize exceptional payoffs. Inf. Manag., 1–12 (2001a)
35. Watson, H., Ariyachandra, T., Matyska Jr., R.J.: Data warehousing stages of growth. Inf. Syst. Manag. **18**(3), 42–50 (2001)
36. Wells, J.D., Hess, T.J.: Understanding decision-making in data warehousing and related decision support systems: an explanatory study of customer relationship management application. Inf. Res. Manag. J. **15**(4), 16–32 (2002)
37. Abdel Hafez, H.A., Kamel, S.: Web based data warehouse in the Egyptian cabinet information and decision support center. In: Decision Support in an Uncertain and Complex World: The IFIP T C8/WG8.3 International Conference, pp. 402–409 (2004)
38. Rubin, D.L., Desser, T.S.: A data warehouse for integrating radiologic and pathologic data. J. Am. Coll. Radiol. **5**(3), 210–217 (2008)
39. Mawilmada, P.K.: Impact of a data warehouse model for improved decision-making process in healthcare. Masters by Research thesis, Queensland University of Technology, October 2011
40. Ojeda-Castro, Á., Ramaswamy, M., Rivera-Collazo, Á., Jumah, A.: Critical factors for successful implementation of data warehouses. Issues Inf. Syst. **12**(1), 88–96 (2011)
41. Alshboul, R.: Data warehouse explorative study. Appl. Math. Sci. **6**(61), 3015–3024 (2012)
42. Shen, L., Liu, S., Chen, S., Wang, X.: The application research of OLAP in police intelligence decision system. Procedia Eng. **29**, 397–402 (2012)
43. Shams, K., Farishta, M.: Data warehousing: toward knowledge management. Top. Health Inf. Manag. **21**(3), 24–32 (2001)
44. Chenoweth, T., Corral, K., Demirkan, H.: Seven Key Interventions for data warehouse success. Commun. ACM **49**(1), 115–119 (2006)
45. DeLone, W.H., McLean, E.R.: Information systems success: the quest for the dependent variable. Inf. Syst. Res. **3**(1), 60–95 (1992)
46. Kumar, R.L.: Justifying data warehousing investments. In: Becke, S. (ed.) Data Warehousing and Web Engineering, pp. 100–102 (2002)
47. Watson, H.J., Gerard, J.G., Gonzalez, L.E., Haywood, M.E., Fenton, D.: Data warehousing failures: case studies and findings. J. Data Warehouse **4**(1), 44–55 (1999)
48. Sammon, D., Adam, F., Carton, F.: Benefit realisation through ERP: the re-emergence of data warehousing. Electron. J. Inf. Syst. Eval. **6**(2), 155–164 (2003)
49. Aguila, M.D., Felber, E.: Data warehouses and evidence-based dental insurance benefits. J. Evid. Based Dent. Pract. **4**(1), 113–119 (2004)
50. Devlin, B.: Data Warehouse from Architecture to Implementation. Addison Wesley Longman Inc., Reading (1997)
51. Griffin, R.K.: Data warehousing. Cornell Hotel Restaurant Adm. Q. **39**(4), 28–40 (1998)
52. Wilkoff, N., Pohlmann, T., Hudson, R., Lambert, N.: The State of Technology Adoption, Business Technographics North America, Forrester Research Inc., 5 May 2004
53. Agosta, L.: Hub-and-Spoke architecture favored. DM Rev. **15**(3), 14–63 (2005)

54. Hwang, H.-G., Ku, C.-Y., Yen, D.C., Cheng, C.-C.: Critical factors influencing the adoption of data warehouse technology: a study of the banking industry in Taiwan. Decis. Support Syst. **37**, 1–21 (2004)

55. Hong, S., Katerattanakul, P., Hong, S.-K., Cao, Q.: Usage and perceived impact of data warehouses: a study in Korean financial companies. Int. J. Inf. Technol. Decis. Making **5**(2), 297–315 (2006)

56. Agosta, L., Andrews, M., Ritzmann, M.: The Data Warehouse Satisfaction Survey, Part 1: The Number One Complaint About Data Warehousing. Information Management Special Reports, 2 October 2007

57. Merritt, K.L.: User satisfaction in data warehousing: an empirical investigation of salient variables. Issues Inf. Syst. **9**(2), 500–508 (2008)

58. Wixom, B.H., Watson, H.J., Reynolds, A.M., Hoffer, J.A.: Continental airlines continues to soar with business intelligence. Inf. Syst. Manag. **25**, 102–112 (2008)

59. Almabhouh, A., Saleh, A.R., Azizah, A.: Examining the influence of relationship quality on data warehouse success. Int. J. Model. Optim. **1**(5), 402–409 (2011)

60. Lodico, M.G., Spaulding, D.T., Voegtle, K.H.: Methods in Educational Research from Theory to Practice. Wiley, San Francisco (2006)

61. Al-Allak, B.: Evaluating the adoption and use of internet-based marketing information systems to improve marketing intelligence (the case of tourism SMEs in Jordan). Int. J. Mark. Stud. **2**(2), 87–101 (2010)

62. Al Khattab, A.: The role of corporate risk managers in country risk management: a survey of Jordanian multinational enterprises. Int. J. Bus. Manag. **6**(1), 274–282 (2011)

63. Chongruksut, W.: The adoption of activity-based costing in Thailand, doctoral thesis, Faculty of Business and Law, Victoria University (2002)

64. Lee, Y.W., Strong, D.M., Kahn, B.K., Wang, R.Y.: AIMQ: a methodology for information quality assessment. Inf. Manag. **40**, 133–146 (2002)

65. AbuShanab, E., Pearson, J.M., Setterstrom, A.J.: Internet banking and customers' acceptance in Jordan: the unified model's perspective. Commun. Assoc. Inf. Syst. **26**, Article 23, 493–524 (2010)

66. Alhawary, F.A., Abommman, A.H.: Measuring the effect of academic satisfaction on multi-dimensional commitment: a case study of applied science private university in Jordan. Int. Bus. Res. **4**(2), 153–160 (2011)

67. Maqbool-ur-Rehman, S.: Which management accounting techniques influence profitability in the manufacturing sector of Pakistan? Pak. Bus. Rev., 53–105 (2011)

68. Ramakrishnan, T., Jones, M.C., Sidorova, A.: Factors influencing business intelligence (BI) data collection strategies: an empirical investigation. Decis. Support Syst. **52**, 486–496 (2012)

69. Saban, M., Efeoğlu, Z.: An examination of the effects of information technology on managerial accounting in the Turkish iron and Steel Industry. Int. J. Bus. Soc. Sci. **3**(12), 105–117 (2012)

70. Ebimobowei, A., Binaebi, B.: Analysis of factors influencing activity-based costing applications in the hospitality industry in Yenagoa, Nigeria. Asian J. Bus. Manag. **5**(3), 284–290 (2013)

71. Hardan, A.S., Shatnawi, T.M.: Impact of applying the ABC on improving the financial performance in telecom companies. Int. J. Bus. Manag. **8**(12), 48–61 (2013)

72. Gerber, S.B., Finn, K.V.: Using SPSS For Windows Data Analysis and Graphics, 2nd edn. Springer Science Business Media, Inc. (2005)

73. Pallant, J.F.: SPSS survival manual: a step by step guide to data analysis using SPSS for Windows (Version 12), 2nd edn. Allen & Unwin, Australia (2005)
74. George, D., Mallery, P.: SPSS for Windows Step-by-Step: A Simple Guide and Reference, 14.0 update, 7th edn. Allyn & Bacon (2006)

Knowledge Extraction from Professional E-mails

Nada Matta[✉], Hassan Atifi, and François Rauscher

Institute ICD/Tech-CICO, University of Technology of Troyes,
12 rue Marie Curie, CS 42060, 10010 Troyes Cedex, France
{nada.matta,hassan.atifi,francois.rauscher}@utt.fr

Abstract. Some professional e-mails contain knowledge about how actor face problem in order to realize projects. This type of knowledge is produced in cooperative activity. Representing project knowledge leads to structure link between coordination, cooperative decision-making and communication. The main objective of our work is to extract knowledge from daily work. So the main questions of our research are:

- Can we extract knowledge from professional e-mails?
- If so, which type of knowledge can be represented?
- How to link this knowledge to project memory?

We present in this paper our first work in this aim. Our hypothesis is tested on a software development application.

Keywords: Knowledge engineering · Knowledge management · Project memory · Traceability · Professional e-mails · Pragmatics analysis

1 Introduction

Currently designers use knowledge learned from past projects in order to deal with new ones. They reuse design rationale memory to face new problems. Knowledge Management provides techniques to enhance learning from the past [12]. Their approaches aim at making explicit the problem solving process in an organization. Their techniques are inherited mainly from knowledge engineering. So, we find in these approaches in one hand, models representing tasks, manipulated concepts and problem solving strategies, and in the other hand, methods to extract and represent knowledge. We note for instance MASK [14, 25] and REX [22] methods. These methods are used mainly to extract expertise knowledge and allow defining profession memories.

But, design projects involve several actors from different fields. These actors produce knowledge when interacting together and take collaborative decisions. So, it is important to also tackle this type in knowledge, which is generally volatile.

We deal, in our approach with this type of knowledge, called Project memory [24]. Project memory must represent organizational and cooperative dimension of knowledge. Current techniques used in Knowledge management, based on expert interviews are not adapted to extract these dimensions of knowledge. To tackle knowledge produced in collaborative activity, we need techniques that help to extract knowledge from

© IFIP International Federation for Information Processing 2015
E. Mercier-Laurent et al. (Eds.): AI4KM 2014, IFIP AICT 469, pp. 43–57, 2015.
DOI: 10.1007/978-3-319-28868-0_3

daily work. In this paper, we present a technique that help to extract knowledge from professional e-mails. The presented approach allows structuring extracted concepts and linking them to the project context. We use pragmatics analysis and knowledge engineering techniques for this aim.

Problem solving plays a central role in design projects. At first in the initial analysis that leads to specification documents, and then during the life-cycle of the project. This is especially true in software design projects if the development follows an agile method, with several roundtrips from design to delivery.

Project Memory focus on keeping "project definition, activities, history and results "as wrote Tourtier in [28]. Problem Solving is an essential part of design rationale as it tackles with problem definition, suggestions, and decision.

Face to face meetings are commonly used in office to manage projects and do collaborative work, but other mediated communications usages are increasing like phone, emails or instant messaging. However, information in emails is volatile, unstructured and distributed among users email boxes, making it difficult to trace and keep for corporate memory.

2   Project Memory

A project memory is generally described as "the history of a project and the experience gained during the realization of a project" [24].

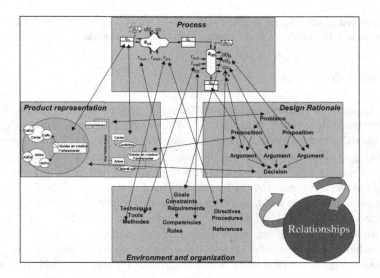

Fig. 1. Project memory

It must consider mainly (Fig. 1):

- The project organization: different participants, their competences, their organization in sub-teams, the tasks, which are assigned to each participant, etc.

- The reference frames (rules, methods, laws, ...) used in the various stages of the project.
- The realization of the project: the potential problem solving, the evaluation of the solutions as well as the management of the incidents met.
- The decision making process: the negotiation strategy, which guides the making of the decisions as well as the results of the decisions.

Often, there are interdependence relations among the various elements of a project memory. Through the analysis of these relations, it is possible to make explicit and relevance of the knowledge used in the realization of the project. The traceability of this type of memory can be guided by design rationale studies and by knowledge engineering techniques.

The problem solving is part of the project design rationale memory [24]. Some technics from knowledge (e.g. REX) aims at knowledge capitalization, but others are more oriented on the traceability of the design rationale. A clear representation of the context and design rationale can be found in [3] and is presented in Fig. 2.

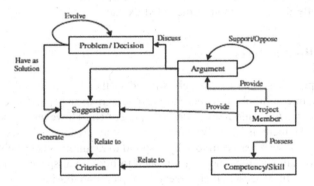

Fig. 2. Problem solving and design rationale in context.

3   Problem Solving

Theory of Human Problem Solving was developed from the work of Newell and Simon [26] and provided the basis of much problem solving research. According to Hardin in [17], "Any problem has at least three components: givens, goal and operation". This general definition from problem solving theory is bringing keys elements into light:

- Givens: information and facts presenting context;
- Goal: desired end state;
- Operations: actions to be performed to reach end state.

In our present study, related to software development, we will focus more on givens and goal, i.e. the "problem recognition" part, the operations being part of the solution. When designing software complex problem solving arise more easily, because the tasks are abstracts and often not well structured as opposed as designing a real world artifact.

Problem solving in professional project aims at transforming knowledge into business value [16]. This usually involve two types of knowledge: declarative (about facts, events, and objects) and procedural (knowing how to do things).

4   Software Development Process

Methodologies in software development evolved quickly in the last two decades from classic waterfall model to Extreme Programming and Agile methods [2].

Agile development is iterative and incremental with continuous delivery. As a side effect roundtrips between product-owner (contractor) and product-manager (developer) are more frequent, leading to increased communication and collaborative work. Typically a software design cycle in agile is divided into sprints, where the product-owner meet the product-manager (developer) and validates recent features, raises issues and express new needs. Problem Solving sequences happen on weekly (sometime daily) basis and implies all the actors of the project, not only the development team. With the new means of communication and project management methods like Kanban [19], this occurs frequently through computer mediated exchanges.

5   Pragmatic Analysis

The act of request has been extensively studied in the field of theoretical linguistics (Searle 1969), intercultural and inter-language pragmatics [4], NLP community on automated speech act identification in emails [7, 22] etc. However, as pointed by De Felice et al. [11] there is very little work concerned with data other than spoken language and few researches seem to fully respond to requirements of being sufficiently general, non-domain specific, and easily related to traditional speech acts. In addition, few researchers have focused their research requests in business written discourse (workplace email communication). Lampert et al. [20] try to create tools that assist email users to identify and manage requests contained in incoming and outgoing email. Atifi et al. [1] analyze email effectiveness from the professional's point of view by mixing two kinds of analysis: a content analysis of interviews of professionals and a pragmatic and conversational analysis of emails. De Felice et al. [11] propose a global classification scheme for annotating speech acts in a business email corpus based on traditional speech act theory described by Austin and Searle [27].

6   Related Works on Email Analysis

Several approaches study how to analyze e-mails as a specific discourse. We note for instance, tagging work [29], in which Yelati presents techniques that help to identify topics in e-mails, or the use of zoning segmentation in [21]. Other works use natural language processing in order to identify messages concerning tasks and commitment [18]. They parse verbs and sentences in order to identify tasks and they track messages between senders and receivers.

Even there is lot of work on pragmatics, which study dialogue and distinguish techniques in order to identify speech intention (Patient/doctor dialogue analysis [17]), coding dialogue scheme [8], etc. Pragmatics analysis of e-mails uses only some of these methods like ngrams analysis by Carvalho in [8], Verbal Response Mode scheme by Lampert in [21] or a custom coding scheme like De Felice et al. [11].

Techniques studying e-mails, often do not consider the context of discussions, which is important to identify speech intention. We deal with our work with professional e-mails, extracting from projects. So, we mix pragmatics analysis and topic parsing and we link this type of analysis to project context (skill and role of messages senders and receivers, project phases, and deliverables, etc.) in order to keep track of speech intention. As pragmatics analysis shows, there is not only one grid to analyze different types of speech intention. In project memory, we look for problem solving, design rationale, coordination, etc. In this study, we focus on problem solving and we build an analysis grid for this purpose.

7 Project Knowledge Extraction from Emails

The main objective of our work is to extract knowledge from daily work. So the main questions of our research are:

- Can we extract knowledge from professional e-mails?
- If so, which type of knowledge can be represented?
- How to link this knowledge to project memory?

To answer these questions, we analyze professional e-mails related to projects. In last studies, we identify a structure to analysis coordination messages [23]. Based on pragmatics analysis, we defined a grid to structure coordination messages based on the main act to do (inform, request, describe, etc.) and the objects of coordination (task, role, product, etc.). In this paper, we will go ahead and define an approach that helps to extract knowledge from professional e-mails. So, we identify firstly step by step how to isolate important messages and how to analyze them. Knowledge from e-mails, as knowledge produced in daily work, cannot be very structured. It is related closely to context. In our work, we focus on knowledge produced during project realization. We will show in our method how information from project organization help in e-mails knowledge extraction.

7.1 Classification of E-mails

Firstly, we have to identify important messages (Fig. 3). For that, we have to gather messages in subjects. Then, we can identify the volume of messages related to each subject. Then we analyze only messages that heave more than 4 answers; we believe that knowledge can be extracted based on interaction. Finally, we link the messages to be analyzed to project phases.

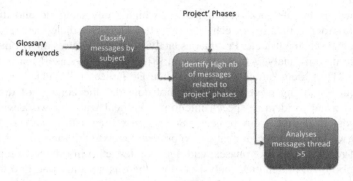

Fig. 3. First e-mails analysis

7.2   Messages Analysis

For each message thread (message and answers), we identify (Fig. 4):

- Information to be linked to organization:
- Authors, To whom, In Copy.
- Information about phases:
- Date and hour of messages and answers.
- Information about product:
- Topic and joined files.
- Information about message intention:
- Main speech act and intention of message.

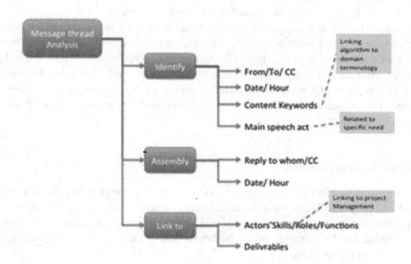

Fig. 4. Analysis of messages

By linking messages to project organization, we help in making sense of interactions between actors. In fact, the role and skill of messages' senders and receivers help to analyze the role of the message in problem solving and the nature of the content (solution answering a problem, proposition discussions, coordination messages, etc.). In the same way, linking messages to phases help to identify main problems to deal in each phase of the same type of projects.

As first work, we focus our speech act analysis on problem solving by identifying request and solution. So, we identify first speech acts that help to localize a request in a message (Fig. 4). Then, we study the organization of related messages thread in order to identify the solution proposed (if it exists) to the request. Our analysis is based first on pragmatics in order to characterize request speech act, and that by identifying request verbs and forms. In the present study we limited our research to the analysis of the act of requesting in problem solving sequences.

From a pragmatic point of view, a request is a directive speech act whose purpose is to get the hearer to do something in circumstances in which it is not obvious that he/she will perform the action in the normal course of events [27]. By introducing a request, the speaker believes that the hearer is able to perform an action. Request strategies are divided into two types according to the level of interpretation (on the part of the hearer) needed to understand the utterance as a request. The two types of requests include direct request and indirect requests. The request can be emphasized either projecting to: 1- the speaker (Can I do X?) or 2- the hearer (Can you do X?). A direct request may be use an imperative, a performativity, obligations and want or need statements.

An Indirect request may use query questions about ability, willingness, and capacity etc. of the hearer to do the action or use statements about the willingness (desire) of the speaker to see the hearer doing x. At last, for us, a grammatical utterance corresponds to only one speech act as in Table 1.

Table 1. Grid of request speech act

Request form	Linguistic form	Examples
Direct request	Imperative	Do x
	Performative	I am asking you to do x.
	Want or Need statements.	I need/want you to do x
	Obligation statements	You have to do x
	Query questions about ability of the hearer to do X	Can you do x?
		Could you do x?
	Query questions about Willingness of the Hearer to do X	Would you like to do x?
	Statements about the willingness (desire) of the speaker	I would like if you can do X
		I would appreciate if you can do X

Then, we complete our analysis by from one side identifying answers verbs and from another side, linking answers to actors' role and skills and also joining files. The date of answers can be an indicator of several elements in the organizations: engagement, difficulty of time spending of solution, stress and multi-responsibilities, etc. We aim at analyzing in the future the frequency of answers.

8   Example

8.1   Example Description

INFOPRO Business Publishing Company asked a software Company to develop a workflow tool that helps journalists to edit their articles and to follow the modification of the journal. Due to geographical constraints, nearly all the communications and negotiations during specification, implementation, tests and delivery were done through email. The development method was mixed agile, with weekly deliveries after initial analysis. The period of the project was more than one year. In this project, the actors were:

- SRA: an editing responsible (skill: law and management, Role: Contractor).
- JBJ: Information System Manager (Skill: Information system, Role: Contractor).
- FX: Information System Developer (Skill: Software Engineering, Role: Development manager).
- CV: Prototyping (Skill: Human Machine Interface, Role: User Interface Modeling).
- RT: Information System Developer (Skill: Software Engineering, Role: Sub-contractor).

Principles phases of the project can be found in Table 2:

Table 2. Phases of the project

	Q1 09	Q2 09	Q3 09	Q4 09	Q1 10	Q2 10	Q3 10
XML Import	▓						
DocumentDatabasespecification and development		▓					
Workflow Specification and development		▓			▓		
User Interface		▓					
Export to magazine and website			▓			▓	
Web service specification and development				▓	▓		
Application test					▓		

8.2   E-mails Analysis

As first step, we identify messages topics based on e-mails subjects. In our project, we identify main discussions topics based on keywords:

- XML: structuration, tag, tree, xsd, dtd, schema.
- BDD: Data base, table, editing part, code part.
- Interface workflow: UI, Workflow, User Interface, login, user management.
- Code: Insurance Code, Legifrance, auto code, vehicle code, mutu code, chapter, article.
- Document: new collection, construction, document.
- Export paper: Indesign, layout, mapping, tag indd, indd template.
- Export site: export web, web tag, web format, dtd web.
- Export Author: word, author, xslt author.
- Services: update legi, Legifrance update, FTP.
- Word: macro word, addin, web service, word 2007, wordlink.

Business emails collected from a project in their raw form are very redundant. In case of multiple replies or forward, several parts of the messages are repeated (e.g. quoted reply content). This occurs typically in long threads, mediated equivalents to spoken conversations, which are especially interesting for our study. Some preprocessing steps have to be performed in order to prepare messages and threads for analysis. We chose a deliberately simple method similar to Carvalho [6]. The steps involved are:

- Remove all previous message text from reply;
- Keep previous message in case of first reply of a thread or forwarded email cause it carries context information;
- Remove signatures and disclaimers when possible (identity of sender and receivers are kept in email metadata).

This leaves us with a corpus of messages and threads without too much duplicated or useless information. For some treatments, the granularity at message level is not sufficient, and it's relevant to split the messages into sentences. Here again, we use a standard approach and split according to punctuation and paragraph signs.

8.3   Frequency

Our corpus represent 3080 messages/14987 sentences in 801 threads between 30 projects actors. Sizes of message are relatively uniform, very long message are not frequent, being not suitable for email efficiency. On the average, threads length is between 2 and 7 messages with some exception at 17 or 21 messages. Usually threads are spread over 3 or 4 days, with higher messages frequency in the beginning.

We identify 10 main actors during this project that account for more than 80 % of the messages. Also the daily frequency in Table 3 show 3 relevant spikes of activity matching critical time of the project: the first delivery and second delivery and a new features addition. We will reduce our investigation to the first spike between 06/2009

Table 3. Daily message frequency.

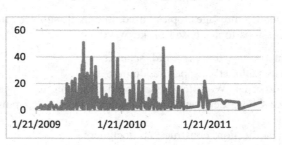

and 09/2009 where a lot of exchanges occurs and focus on long threads showing the presence of a dialog.

As an additional information, it is to be noted that the Development Manager was the one receiving in TO (direct receiver) the higher number of messages, the Chief Editing Manager was the one sending message the most and the Information System Manager was the one receiving the most message in CC. This global numbers are matching their roles in the project, respectively executing, requesting and supervising.

8.4　Topics

We decide to make a very straightforward and knowledge oriented classification of messages and sentences. This steps is necessary in order to assert to deal with messages concerning pure software functionality knowledge and to filter project coordination emails.

Our approach is to create a keywords dictionary for the main topics of the project. This dictionary can be built from the following sources:

- Project phasing and specifications documents;
- an expert;
- domain ontology if available.

As in project memory context, we choose not to rely on statistical NLP clustering like in Cselle's approach [10] but to use existing context knowledge. This dictionary is voluntarily kept simple and have the form:

Topic1: keywords1, keywords2. . . keywordsn.

Using this dictionary we classify messages into weighted topics vector (same technic is applied to sentences for a fine granularity analysis). In order to do that we use a cosine similarity based algorithm. We compute a Lucene [14] ranking between our message and each topics in order to identify main topics of messages (boosting email subject importance compared to email body (Figs. 5 and 6). This give us a topics matrix T where T_{ij} represents weight of topic j in message i.

$$score(q,d) = coord\text{-}factor(q,d) \cdot query\text{-}boost(q) \cdot \frac{V(q) \cdot V(d)}{|V(q)|} \cdot doc\text{-}len\text{-}norm(d) \cdot doc\text{-}boost(d)$$

Fig. 5. Lucene scoring formula

As a side remark, keywords chosen in topics shall not overlap too much to keep the results significant. In Fig. 6, one can notice the amount of emails increased as the project is approaching its first milestone, but the topics are not always directly correlated to the phase of the project. In fact the project team members are often exchanging emails and dealing with problem before the phase really start or after (when the phase is supposed to be finished and some problems remained unsolved).

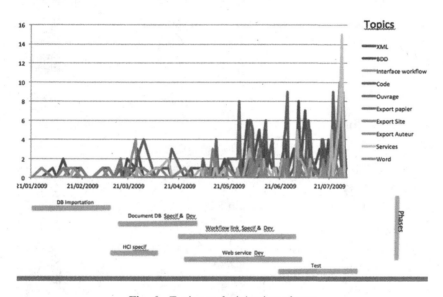

Fig. 6. Topics analysis/project phases

Figure 7 shows first step of analysis of these messages; in which we show senders and receivers and their skills, topics of messages and date of messages. Some patterns of communication are emerging, for instance, the Information System Manager (JBJ) is very often in CC of every message because of its supervisor role.

To analyze messages text, we use pragmatics in order to identify problem and solution discussions. For that, we identify Request messages based on Request speech acts. Then, we identify related answers messages. In these messages we look for sender's skills and joined files. So, we identify for the "Annexes" topic, in which there are 23 messages, related topics are XML and code. Messages were during 12 days from 5th until 17th June. They concern workflow development phase. Based on the Request-Answer grid and role actors, we analyze messages, in order to identify

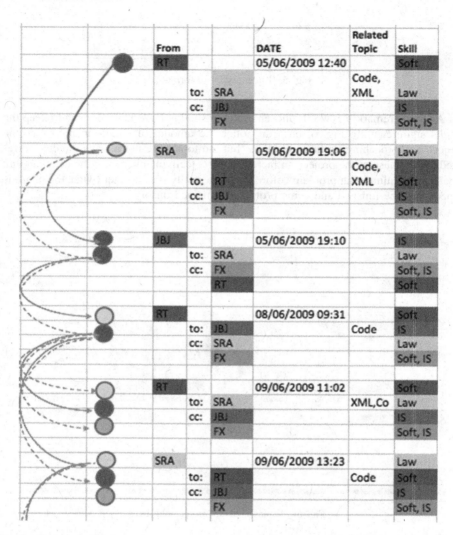

Fig. 7. First analysis of messages: representing of Senders/Recievers/Copy, date and actors role and skill

problem-solving intentions. So, we identify for instance, the problem Insurance Text extraction. SRA; the editing responsible (contractor) asks FX to extract Insurance text in a good format. When FX; the Information System Developer (Development manager) answer him, we suppose that as an answer, based on the role of sender of message and the main topic. We consider also joined files as part of this answer. Figure 8 shows this example.

From			Date	Sentence elements	Related Topic	Function
SRA			2009-06-05 12:40:46.	I put in "Bold", what **I need**:		Request
	to:	FX		1- *Insurances*		
	cc	JBJ		2- Text without tags Text in XML files	Code	
		CV		3- Tag Pb : Text outside tag in XML	XML, Code	
		RT		4- Tag Pb <b> is opened and not closed, as same as, tag is badly formed		
FX			2009-06-05 19:06:34.			Answer
	to:	SRA		1- *Insurances*		
	cc:	JBJ		I propose to convert: Xpress format in XML	XML	
		CV		Beware, the text will contain a lot of error blank, "enter" and image	Code	
		RT		I can transform it on enriched XML	XML	
				It contains a lot of references, so we have to compose with links		

Fig. 8. Example of messages analysis

9 Conclusions

The aim of our study is to identify knowledge from daily work. In this paper, we show that it is possible to study professional e-mails for this aim. We consider e-mails as specific discourse. So we use pragmatics generally used to analyze discourse and to categorize it to identify knowledge from professional e-mails. Our hypothesis is can we identify a grid as guide to analyze professional e-mails? If so, can the result be relevant as project knowledge?

Based on this hypothesis, we know that pragmatics intention must be based to context. So, we consider the project context from different aspect: organization and environment. We believe that this context is very helpful to clarify ambiguity of sentence analysis. We show in the example how sender/receiver role can identify problem-solving answer. Adding this analysis to the identification of keywords of messages, as topics can be a first step, towards a structuring of knowledge: Problem related to a topic, possible answers.

We will continue to validate this work on other type of projects. This work can open to identify other grid analysis like: engagement of actors, design-rationale, coordination [23], etc. Our current work objective is to explore various techniques from machine learning to implement support algorithm for the projection of our features vectors (topics, pragmatics and context) to problem- solving knowledge model. Although related to the works of Cleland-Huang et al. [5] on Requirement traceability in software design, we will focused more on functional testing and design detection.

Finally, this study is a part of our work on project memory: Keeping track and structuring knowledge in daily work realization of project. We developed techniques to extract knowledge from project meetings [13] and to identify occurrences in order to identify concepts in project memory.

References

1. Atifi, H., Gauducheau, N., Marcoccia, M.: The effectiveness of professional emails: representations and communicative practices. In: Proceedings of the 13th Conference of the International Association for Dialogue Analysis, Dialogue and Representation, Montréal (2011)
2. Beck, K., Beedle, M., Van Bennekum, A., Cockburn, A., Cunningham, W., Fowler, M., Thomas, D.: Manifesto for agile software development (2001)
3. Bekhti, S., Matta, N.: Project memory: an approach of modelling and reusing the context and the de design rationale. In: Proceedings of the International Joint of Conferences of Artificial Intelligence (IJCAI 2003), Workshop on Knowledge Management and Organisational Memory, Acapulco (2003)
4. Blum-Kulka, S., House, J., Kasper, G. (eds.): Cross-Cultural Pragmatics: Requests and Apologies. Ablex Publishing, Norwood (1989)
5. Cleland-Huang, J., Settimi, R., Zou, X., Solc, P.: The detection and classification of non-functional requirements with application to early aspects. In: 14th IEEE International Conference on Requirements Engineering, pp 39–48 (2006)
6. Carvalho, V., Cohen, W.: Learning to extract signature and reply lines from email. In: Proceedings of the Conference on Email and Anti-spam, Palo Alto, CA (2004)
7. Carvalho, V., Cohen, W.: On the collective classification of email "speech acts". In: Proceedings of the 28th Annual International ACM SIGIR Conference on Research and Development in Information Retrieval, pp. 345–352. Association for Computing Machinery, New York (2005)
8. Carvalho, V., Cohen, W.: Improving "email speech acts" analysis via n-gram selection. In: Proceedings of the HLT-NAACL 2006 Workshop on Analyzing Conversations in Text and Speech (ACTS 2009), pp. 35–41. Association for Computational Linguistics, Stroudsburg (2006)
9. Core, M.G., Allen, J.: Coding dialogs with the DAMSL annotation scheme. In: AAAI Fall Symposium on Communicative Action in Humans and Machines (1997)

10. Cselle, G., Albrecht, K., Wattenhofer, R.: BuzzTrack: topic detection and tracking in email. In: Proceedings of the 12th International Conference on Intelligent User Interfaces, pp. 190–197 (2007)
11. De Felice, R., Darby, J., Fisher, A., Peplow, D.: A classification scheme for annotating speech acts in a business email corpus. ICAME J. **37**, 71–105 (2013)
12. Dieng, R., Corby, O., Giboin, A., Ribière, M.: Methods and tools for corporate knowledge management. In: Proceedings of the KAW 1998, Banff, Canada (1998)
13. Ducellier, G., Matta, N., Charlot, Y., Tribouillois, F.: Traceability and structuring of cooperative Knowledge in design using PLM. Knowl. Manage. Collab. Spec. Issue Int. J. Knowl. Manage. Res. Pract. **11**(1), 53–61 (2013)
14. Ermine, J.L.: La gestion des connaissances. Hermès Sciences Publications, Paris (2002)
15. Gospodnetic, O., Hatcher, E.: Lucene in Action. Manning Publications, Greenwich (2004)
16. Gray, P.H.: A problem-solving perspective on knowledge management practice. Decis. Support Syst. **31**(1), 87–102 (2001)
17. Hardin, L.E.: Problem solving concepts and theories. J. Vet. Med. Educ. **30**(3), 227–230 (2002)
18. Kalia, K.A.: Identifying Business Tasks and Commitments from Email and Chat Conversations. Technical report, HP Labs (2013)
19. Ladas, C.: Scrumban-essays on kanban systems for lean software development (2009). http://Lulu.Com
20. Lampert, A., Dale, R., Paris, C.: Detecting emails containing requests for action. In: Proceedings of the Human Language Technologies: The 2010 Annual Conference of the North American Chapter of the Association for Computational Linguistics (HLT-NAACL), pp. 984–992. Association for Computational Linguistics (2010)
21. Lampert, A., Dale, R., Paris, C.: Classifying speech acts using Verbal Response Modes. In: Proceedings of the 2006 Australasian Language Technology Workshop (ALTW 2006), pp. 34–31 (2006)
22. Malvache, P., Prieur, P.: Mastering corporate experience with the REX method. In: Proceedings of the International Symposium on Management of Industrial and Corporate Knowledge (ISMICK 1993), Compiegne (1993)
23. Matta, N., Atifi, H., Sediri, M. Sagdal, M.: Analysis of interactions on coordination for design projects. In: IEEE Proceedings of the 5th International Conference on Signal-Image Technology and Internet Based Systems, Kula Lumpur (2010)
24. Matta, N., Ribière, M., Corby, O., Lewkowicz, M., Zacklad, M.: Project Memory in Design, Industrial Knowledge Management - A Micro Level Approach. Springer, Verlag (2000)
25. Matta, N., Ermine, J.-L., Aubertin, G., Trivin, J.-Y.: Knowledge capitalization with a knowledge engineering approach: the MASK method. In: Dieng-Kuntz, R., Matta, N. (eds.) Knowledge Management and Organizational Memories, pp. 17–28. Kluwer Academic Publishers, New York (2002)
26. Newell, A., Simon, H.A.: Human Problem Solving. Prentice-Hall, Inc., Englewood Cliffs (1972)
27. Searle, J.R.: Speech Acts: An Essay in the Philosophy of Language. Cambridge University Press, Cambridge (1969)
28. Tourtier, P.A.: Analyse préliminaire des métiers et de leurs interactions. In: Rapport intermédiaire, project GENIE, INRIA-Dassault-Aviation (1995)
29. Yelati, S., Sangal, R.: Novel approach for tagging of discourse segments in help-desk e-mails. In: 2011 IEEE/WIC/ACM International Conference on Web Intelligence and Intelligent Agent Technology (WI-IAT), vol. 3, pp. 369–372 (2011)

Challenges for Knowledge Management in the Context of IT Global Sourcing Models Implementation

Kazimierz Perechuda and Małgorzata Sobińska[(✉)]

Wrocław University of Economics, Wrocław, Poland
{kazimierz.perechuda,malgorzata.sobinska}@ue.wroc.pl

Abstract. The article gives a literature overview of the current challenges connected with the implementation of the newest IT sourcing models. In the dynamic environment, organizations are required to build their competitive advantage not only on their own resources, but also on resources commissioned from external providers, accessed through various forms of sourcing, including the sourcing of IT services. This paper presents chosen aspects of knowledge management and knowledge and information security, in the context of IT sourcing models implementation. IT sourcing solutions are presented, as employed by modern companies, together with potential benefits offered. The main focus is put on the determination of the most important risks involved in knowledge sharing in IT sourcing relations, as well as minimization and reduction of such risks, with particular attention to the newest trend in IT sourcing - cloud computing services on offer.

Keywords: Management · Knowledge management · IT sourcing models · Cloud computing · Information security

1 Introduction

Businesses are on the lookout for newer and more innovative ways to enhance competitiveness and get ahead of the growth curve. A new generation of advanced technologies – social, mobility, analytics and cloud – have taken the center-stage, promising to transform enterprises and help them do business better. Enterprises that embrace these technologies would be able to seamlessly redesign their business models, strategy, operations and processes to meet the new customer demands.

A review of the wider outsourcing literature (e.g. [1, 7, 14, 15]) provides a basis to the claim, that as outsourcing spend increases, the alignment of business and sourcing strategy becomes a key issue and needs from CEO and business executive involvement in outsourcing objectives, relationships and implementation. The authors suggest a range of key issues that could be usefully researched in the context of IT sourcing implementation and functioning of IT sourcing relation.

In this article the attention will be paid to the knowledge and information management in IT sourcing relation. The main objective is to analyze the risks and opportunities of knowledge and information sharing in IT sourcing cooperation.

© IFIP International Federation for Information Processing 2015
E. Mercier-Laurent et al. (Eds.): AI4KM 2014, IFIP AICT 469, pp. 58–74, 2015.
DOI: 10.1007/978-3-319-28868-0_4

In authors opinion, the aspects of knowledge management in outsourcing relationship is relatively underrepresented in professional literature and too often underappreciated in practice. It can lead to a variety of negative effects, such as organizational problems, limited communication with service suppliers, low quality of services rendered, mounting costs as well as growing barriers to exit from the outsourcing deal.

The IT departments of modern organizations are still more and more reliant on the services of external providers and suppliers of hardware, software, telecommunication, cloud computing resources, etc. By end of 2013 global outsourcing contract value for business and IT services was about \$US648 billion (BPO \$304b., ITO \$344b.), and by the end of 2014 exceeded \$US700 billion. On some estimates the market will see 4.8 % compound annual growth through to end of 2018 as more is outsourced, and new service lines and delivery locations are added. Within this, offshore outsourcing exceeded \$100 billion in revenues in 2013 and we estimate it to grow at 8–12 % per year from 2013 to 2018. At the same time, despite the increased popularity of IT sourcing, the satisfaction from this type of business model of IT management remains at a relatively low level [12].

One of the objectives of the pilot study of Willcocks and Sobińska was to provide answer to the following question: "Which IT sourcing models are, in the respondents' opinion, burdened with excessive risk?" [12].

When asked about the evaluation of risk involved in various sourcing models, the respondent companies pointed out the elevated risks in offshoring (54 %) and cloud computing (25 %). Those two IT sourcing models are decidedly less popular among Polish companies as at 2014, but the interest in cloud computing solutions is expected to grow in the near future and, with maturity, cloud sourcing emerges as quite attractive for respondent Polish companies.

The reluctance to share information and knowledge, as well as the lack of trust (30 % of the responses) were high on the list of risks materializing reported by companies (see Fig. 1).

The business models employed by modern enterprises are increasingly more involved in problems related to the security and protection of information, data, and knowledge, particularly of the classified and undisclosed type.

In a sense, these business models can be viewed as based on knowledge and security. The classified knowledge (technical, technological, design, logistic, etc.) is one of the core competences of large network corporations, such as Renault, Mazda, Opel, Toyota, Deutsche Bank, and others.

The network structure of large corporations, while designed to provide competitive advantage in two areas, namely:

- outsourcing of ancillary functions, support functions, and even primary functions (as in the case of Opel assembly factory in Gliwice),
- centralized investment in R+D and new technologies (patents, inventions, improvements, copyrights),

may also increase the risk of uncontrolled 'leakage' of key undisclosed knowledge (technical, design, technological, financial, trade, etc.) to market competitors. This is a direct result of the increased access to core corporate knowledge offered to cooperating entities.

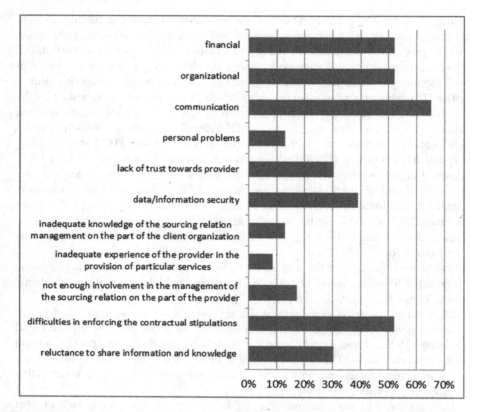

Fig. 1. The most important risk factors materializing in the IT sourcing process (Source: [12]).

The most important aspect of this process is the natural outflow of hot knowledge, resulting from transmigration of knowledge agents (managers and long-term employees with unique competences and experience), in both the networked and non-networked systems.

2 Methodology

This paper combines literature review with the authors comments. It presents a wider look on the knowledge management and security in IT sourcing relationship, that was partly discussed in the paper "Information security in IT global sourcing models" [11].

3 New Global Sourcing Models of Business

In the modern, 'flat' model of economy, networked enterprises build their competitive advantage through careful selection of sourcing agents. One of the most important criteria for such selection is the perceived level of security with respect to uncontrolled and undesired outflow of data, information, and knowledge from organizations to other entities outside their network structure.

Sadly, this particular criterion is rarely perceived as mission-critical. Companies tend to prioritize the aspect of compatibility between core competences of the potential sourcing partner with key competences and resources of the mother company. The increased asymmetry of key competences between sourcing partners may result in the following trends (Fig. 2):

- departure (short-term contracts, rapid capturing of the partner's know-how),
- unification (strengthening the cooperation, balancing the symmetry of undisclosed knowledge, participation in future projects).

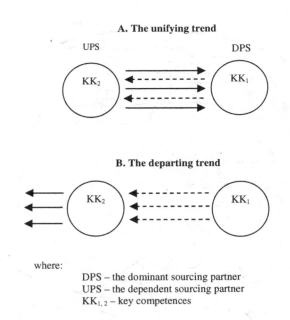

A. The unifying trend

B. The departing trend

where:

DPS – the dominant sourcing partner
UPS – the dependent sourcing partner
$KK_{1,2}$ – key competences

Fig. 2. Trends in sourcing cooperation (source: own research).

New needs of enterprises result in the emergence of new types of global sourcing models, where sourcing can be defined "as the act through which work is contracted or delegated to an external or internal entity that could be physically located anywhere" [1, p. 2]. Sourcing can also be defined as a comprehensive organizational strategy for distribution of business processes and other functional areas of the enterprise among cooperating partners. For the purpose of this study, sourcing is defined as a notion of superior level to the notions of outsourcing and insourcing [2, p. 17]. The main differences between sourcing models involve determination of the following factors:

- is the sourced function delegated to a dependent entity or an independent external supplier (or both), and
- is the sourced function performed on-site or off-site, is it performed onshore, nearshore (in a neighboring country) or offshore (in a remote location) [1, p. 25].

A business model of IT sourcing may comprise the following types/models of sourcing cooperation/relation (own research, based on: [6, pp. 6–16]; [1, pp. 26–42]; [4, p. 1300]; [5]: facility management, selective outsourcing, tactical outsourcing, transformational outsourcing, transitional outsourcing, Business Process Outsourcing, joint ventures, benefit-based relationships, insourcing (staff augmentation), offshore outsourcing (foreign supplier), nearshore outsourcing (foreign supplier), onshore outsourcing (domestic supplier, "rural sourcing"), cosourcing, shared services, captive models and models based on Internet: cloud computing, software as a service, crowdsourcing and microsourcing. Figure 3 presents a graphic representation of a general model of IT sourcing.

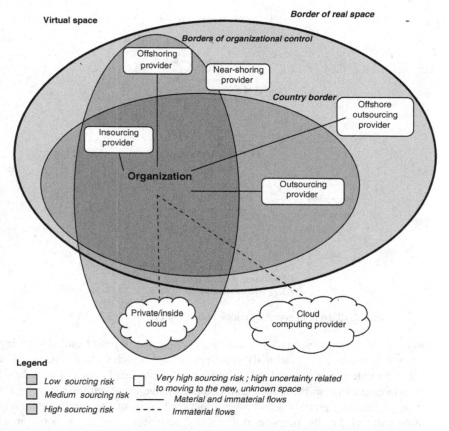

Fig. 3. A general model of IT sourcing (source: [5]).

A large number of modern organizations operate simultaneously in two business areas: the real and the virtual. The greater the availability of resources, the greater the potential impact. The more we expand the range of the network (new locations where functions/processes/services of an organization are implemented; greater number of sourcing providers; new areas of service delivery such as cloud computing), the greater the potential opportunities, but also the greater is the risk involved.

A decision to implement a particular sourcing model may be influenced by the following factors (own research, based [6, p. 4]:

- the global skills shortage,
- a more mobile workforce,
- the mounting cost of in-house developed software,
- the need to move fast, rapidly adopting new technologies and speeding up system development,
- the explosion of Internet technologies and services requiring a wide range of new skills and investments.

For the purpose of this study, the authors focus on the presentation and analysis of one of the most popular Web-based sourcing models – the cloud computing – without going into detailed analysis of other IT sourcing models in use.

Cloud computing allows users to access technology-enabled services on the Web, without having to know or understand the technology infrastructure that supports them. Nor do they have much control over it. It is an innovative way to boost capacity and add capabilities in computing without spending money on new infrastructure, training new personnel or licensing software.

There are four basic types of clouds: private clouds (operated solely for the use of a single organization), community clouds (operated for a specific group that shares infrastructure), public clouds (which use cloud infrastructure available over a public network) and hybrid clouds (which combine the infrastructure of two or more clouds - public, community, and private).

The increased risk of cloud computing projects has to do with opening up the organization to a whole new space, which is not yet completely examined and "protected". The range of potential risk scenarios is impossible to predict at this moment, since they have yet to be observed in organizational practice. At the same time, the output and the use of the "new space-clouds" can increase the potential added value of using this type of sourcing, compared to more classical forms, such as the generic outsourcing and offshoring models.

New forms of contracting, and – consequently – new resource acquisition methods are required to help modern organizations survive in this age of innovation and strong competition. However, it should be noted that those new solutions, as any new ideas, come with new risks and new demands for management. Some of those risks with the principles and methods conducive to their reduction will be discussed in the following sections of the paper.

4   Knowledge Management in IT Sourcing Relationship

Using the services of external provider can become a driver of cultural change processes of an organization and encourage organizational renewal. However, a prerequisite for such a change is the full commitment of management staff. An outsourcing provider can play a dual role in the process of change - on the one hand he can perform professional services innovation and integration, on the other hand he has the tools to carry out organizational and cultural change throughout the organization [2, p. 36].

One can characterize the potential and risks of IT outsourcing in the context of knowledge management as follows:

- Outsourcing (including offshore outsourcing) is conducive to broaden the organizational and technological knowledge of an organization. Specialized IT service providers provide an access to the expertise and skills of experts, which the organization does not have, and that is hard to find in the local market. Planning outsourcing model implementation forces deepening and formalization of knowledge about the organization - the evaluation of its key resources, competencies, implemented processes and needs, as well as the environment - the market opportunities in the provision of IT services. Thanks to the preparation and the implementation of the outsourcing contract organization gets a lot of experience on how to manage outsourcing relationship and the problems that may arise in the course of cooperation with external service providers.
- Outsourcing is conducive to establishing informal relations between customer and service providers employees, which facilitates the exchange of information (including know-how), and helps to follow-up activities of the experts.
- Offshore outsourcing of IT services, although associated with a higher risk and communication difficulties is an opportunity not only to gain access to expertise, but also to observe the best practices of global IT services providers on the standardization and formalization of work, thereby ensuring a better control of the implemented services.
- Outsourcing and offshore outsourcing provides new business contacts and experience in the management of this type of projects.
- The risk of outsourcing and offshore outsourcing (also in the context of knowledge management) is much smaller, if it does not apply to areas and functions that are crucial for the functioning of the organization.
- The experience of Indian offshoring companies customers give rise to the claim that outsourcing services to the companies in the same industry is very risky. Knowledge, which is sent to the provider can greatly weaken the position of the organization and to strengthen the position of the supplier.

Positive and negative experiences with the use of outsourcing in areas such as e.g. information technology, where appropriate transfer of knowledge between the contract parties (especially in the case of total outsourcing) often determines the success or failure of cooperation, suggest that the model of knowledge management in IT sourcing could not only fill the research gap, but also be used in practice.

The effective management of IT outsourcing cooperation requires qualified personnel with suitable skills and knowledge.

The issue of knowledge management in relationships with external suppliers was discussed by M. Sobińska in her previous publication. The author defined knowledge management in outsourcing projects as *"an identification and coordination of processes that influence: knowledge localization, knowledge development, knowledge exchange, knowledge utilization and preservation of knowledge related to carrying out the outsourcing process, based on the use of properly selected methods and instruments and intended to facilitate completion of the outsourcing goals as well as extending of organizational knowledge"* [13, p. 205].

Figure 4 presents a model of knowledge management processes in IT outsourcing.

From the perspective of knowledge management outsourcing can be defined as (based on [16, p. 394]):

- a way of acquiring expertise and skills that the organization does not possess;
- the form of stabilization of the knowledge related to the functioning of selected areas of an organization;
- a guarantee to keep up with technological development (in this case, an outsourcing contract should include appropriate conditions obligating the provider to continuous development and improvement of the provided services);
- a replacement of the internal know-how with the same kind of knowledge from the outside.

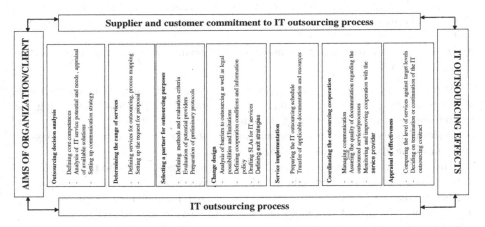

Fig. 4. Model of knowledge management processes for IT outsourcing (Source: based on [13, p. 212]).

Knowledge management in IT outsourcing relationship should enable creation and utilization of intellectual capital, that, in Willcocks opinion, should be generated through the interplay between such essential elements as [1, p. 111–112]:

- structural capital (that refers to the codified bodies of semipermanent knowledge that can be transferred and the tools that augment the body of knowledge by bringing relevant data or expertise to people),
- human capital (which represents the capabilities of individuals to provide solutions to customers),
- customers capital (that is linked to shared knowledge, or the value of an organization's relationships with the people with whom it does business)
- and social capital (for example: trust, loyalty and reciprocity within a community-the values created from social networks), which helps bring these elements together.

Outsourcing implementation frequently disrupts and reduces social capital by removing people, systems, and organizational knowledge from the client organization. It should be paid more attention to strengthening the social capital as it can have a considerable impact on effective knowledge transfer between outsourcing parties. Social capital allows outsourcing partners reduce for example cultural barriers, understand common goals and strengthen network stability and ties.

Modeling of knowledge management processes in outsourcing relationships can significantly contribute to:

- reducing the difficulties of communication with an external provider,
- improve the process of IT sourcing implementation,
- better protection of organizational knowledge,
- effective information and knowledge exchange in both customer-supplier and supplier-customer relationship,
- the real development of organizational knowledge during the outsourcing contract (both organizational knowledge about project management and expertise (which is especially important if the organization after the contract termination wants to return to providing services internally),
- improvement of the outsourcing process for achieving the strategic objectives of the organization.

Still more and more outsourcing clients believe that part of an outsourcing transaction is access to innovative thinking. At the same time the teams managing the relationship with the supplier focus on reducing costs and effectively block any proposal (from the providers) to amend and modify the outsourced services not passing them on to the end users and top management.

Therefore it is quite important for the client organization to give to the provider the right to make "valuable" proposals (which could have a significant positive impact on the functioning of the organization) directly to the Executive Director or/and top management. The provider should also be able to provide suggestions concerning the activities/operations that are out of the contract, if they were to foster innovation [2, pp. 165–166].

The need for innovation resulting from changing business conditions, economic slowdown and globalization and technologization of services delivery enforces a new look at sourcing relationships and the transition from outsourcing management to cooperation for innovation. This change is realistic but requiring from the engaged parties: high maturity, mutual trust, high competence, commitment and flexibility.

These actions resulting from the proper approach to knowledge management in relationships with the IT services providers would gradually move the organizations from basic services sourcing model towards innovative cooperation.

One of the modules of knowledge management in outsourcing relationships, especially in the area of IT should be the information and data security management, that will be discussed in the next chapter. It could be particularly important for the implementation of cloud solutions, especially in cooperation with foreign service providers.

5 Information and Knowledge Security in the IT Sourcing Models

Information security is one of the key factors to be taken into account in the context of sourcing decisions, particularly those which involve cooperation with external partners. Potential contractors may operate from remote locations, often in diverse cultural, political, social, legal, and other settings. The problem of information security is also reported as one of the main concerns for potential cloud computing users and clients.

According to K. Liderman, information security comprises all forms (also verbal) associated with the exchange, storage, and processing of information. It represents the risks involved in information resource loss, as well as misinformation resulting from poor quality of information provided. It must be noted that K. Liderman makes a clear distinction between information security and 'security of information'. The latter, a subset of information security, is defined as *"justified trust (e.g. based on risk analysis and the adopted risk management procedures) that the organization will not face potential losses as a result of undesired (accidental or intentional) use of information stored, processed, and transmitted within the system, be it information disclosure, modification, removal, or rendering it unfit for processing purposes"* [3, p. 22]. The problem of information security, despite increased publicity in professional literature and media, is still trivialized by managers of many companies. And, while the managerial personnel declare their knowledge of risks involved in this area (as reported in many studies, see: [8, p. 19], in practice they tend to minimize their efforts with respect to information security. The main reasons for negligence are: the financial outlay, and the lack of competences. For instance, not many companies are able to perceive the risk associated with the fact that company information is no longer contained within the 'company perimeter'. Nowadays, information flows freely, and the risk of disclosure is far more pronounced. The lack of informed approach to the risk management process is a fast lane to disaster, since potential incidents may gravely affect the security or company development. Modern organizations face the serious challenge of implementing effective security strategies, with proper risk management as one of the main elements of the system [4].

In this context, outsourcing may be seen as a chance to improve information security (and the security of IT systems) by transferring the IT security responsibilities to an external provider with proper specialist knowledge and IT technologies. On the other hand, it may just as well add to the risk, by opening up sensitive company resources to external agents. Those resources may (unwillingly or intentionally) be used to detrimental effects.

Since cloud computing has become a hot topic in IT management, it may be useful to address some of the security issues involved in this form of cooperation. In terms of information security, the main difference between traditional IT structures and cloud solutions is the fact that cloud infrastructure is interspersed and shared among many users. In addition, certain features of cloud computing, such as the need for continuous optimization, improved access, balancing the computing load across the nodes, etc., bring additional complications to the risk management process.

At present, three main categories of cloud computing solutions can be distinguished, namely: SaaS - Software as a Service, PaaS - Platform as a Service, and IaaS - Infrastructure as a Service. Factors of potential risk to information security include the following [9, p. 230]:

- IT system's resistance to intrusive attacks from outside,
- Resistance to internal attacks (with users able to access and capture information belonging to other users, by exploiting security holes and other vulnerabilities of the system),
- The security verification and encryption methods in use (whether the access codes and passwords are stored in secure, encrypted form, or stored and transmitted in open text).

Each of the above models of cloud computing (SaaS, PaaS, IaaS) employs a specific set of information security solutions [9, pp. 232–233]:

- In the SaaS model, users rely on service provider in all matters concerning information security. The provider is responsible for restricting access to sensitive information, as well as supplying proper security measures to prevent intrusion or breakdown. The provider is also responsible for all matters concerning access verification and data encryption. However, the user is rarely able to examine the details of the security measures taken, to make sure that they are up to the desired standard of service.
- In the PaaS model, the service provider may choose to grant the software provider some control over the system security (for instance, the software provider may take on the responsibility of providing their own access verification and encryption systems), but any security issues beyond the application level, such as the security of host machines or network, come under the administration of the service provider. The provider may choose to pass on to the user selected details on the security measures in use.
- In the IaaS model, the software producer has a great deal of control over security mechanisms, since the cloud applications are run on virtual machines, independent and separated from virtual machines used by other users. However, applications in this model take longer to develop, and are decidedly more costly.

The majority of cloud computing service providers offer data backup solutions. This aspect is clearly important from the user's viewpoint, but it must be noted that data backup is not a 100 % solution for all security concerns. The SaaS model, in particular, seems the most risky solution in the context of information security. In this model, both the software code and the data being processed are stored remotely (i.e. outside the subscriber's real machine). Consequently, the user has no access to computing operations, and is in no way able to modify it. SaaS offers the potential for server operator to modify the computing software or data processing procedures. Users must upload their data on the server for computing. The result is the spyware effect: the server operator receives user data freely and effortlessly, due to the character of the service rendered, and this gives him the unfair advantage (or even power) over the user. With the IaaS model, on the other hand, it is advisable to refrain from implementing needless functions on virtual machines, as well as making sure that all virtual machine images communicate over encrypted channels, so as to eliminate the risk of data capture on the network infrastructure level.

According to A. Mateos and J. Rosenberg, the security of the cloud computing environment is comparable to that of most internally managed systems, because:

- Most of the potential (and known) risk problems can be eliminated by employing the existing technologies, such as data encryption and virtual local area networks (VLAN), as well as standard tools, such as firewalls and packet filtering (encrypted data stored on cloud may in fact be safer than non-encrypted data used locally);
- Cloud computing solutions may be supplemented by additional controlling and auditing functionality, layered outside the environment of the host. Such a solution offers the user greater security of cloud computing, far better than any locally implemented solution (since the latter require considerable outlay and design expertise);
- Many countries enforce security measures on SaaS service providers, requiring them to restrict transmission of data and other copyright content to the contractor's country of origin ([10], p. 104)

As aptly put by J. Viega, one of the fundamental benefits of a cloud solution is the fact that those structures are fairly unrewarding for willful intruders, since the code – i.e. the most vulnerable element that can be tested for security holes and exploited – is stored on server side, instead of being sent to the client browser ([9], p. 230). Data centralization in cloud structures, as opposed to decentralized distribution of data within the company network, allows for vast reduction of leak risk, since users are less inclined to store sensitive resources on their real machines. Furthermore, the access to a centralized resource storage and actual data use can be monitored more closely.

The concern for security of information exchanged in the course of company relations with external service providers, although well-founded, must be examined against any potential benefits offered by this particular type of business model. And the actual informational risk may be largely minimized by employing proper principles of management with respect to relations with external providers – this applies also to knowledge management.

6 Ways to Reduce the Risk of Knowledge/Information Loss in IT Sourcing Models

K. Liderman believes that information security can be enhanced by employing proper documentation of the security system in use. This task serves the following purposes [3, p. 120]:

- to ensure proper level of protection with respect to information and to those elements of the system which are directly involved in data processing and storing;
- to track (and control) any changes introduced to the system;
- to satisfy the legal requirements that oblige companies to keep and produce on demand certain documents, such as 'security policy guidelines', 'safety instructions and procedures', etc. (the wording used in actual legal standards may vary).

The use of formal documents (information security policies or guidelines) may attest to the company's intent on keeping proper security standards in data protection. It may also help the organization build and maintain trust relations with customers and/or business partners. Lastly, it may also be used to stimulate the involvement of employees in all tasks and procedures related to data/information security.

With respect to basic technical security measures employed for the purpose of maintaining the informational stability of IT and telecommunication systems, Liderman provides the following classification of elements [4, pp. 158–159]:

- data backup procedures;
- provision of independent backup power solutions;
- provision of backup solutions for data processing (or even for running the company business, if necessary), in a reasonably remote location from company HQs;
- doubling the key infrastructure: servers, routers, etc., to serve as backup;
- doubling the information packets;
- providing alternative transmission routes (doubling keys and operators);
- use of verified software offering suitable protection of transmitted and stored data (proper data transmission protocols, software-assisted verification of data integrity based on cryptographic methods, etc.);
- protecting the telecommunication and IT systems from unauthorized access – both physical (access to hardware and technical infrastructure) and logical (access to information resources);
- protection from intentional or accidental exposure to hazards (fire, flooding, strong electromagnetic impulses, etc.).

The most advanced security measures used in cloud computing data centers include (own research, based on [10, pp. 104–114]):

- physical security – modern centers are often located in unassuming locations and buildings (often in residential areas), with good security and skillful use of barriers (also natural). Security services cover both the immediate area and the access to actual data facilities, using modern CCTV solutions, intruder alert systems, etc. Servers are kept in fortified bunkers, protected by 5-level biometric scanners (hand geometry recognition), round-the-clock patrols and traps (caging intruders in case of unauthorized entry). Physical security is solely in the hands of the cloud computing service provider, and the above measures are required for certification purposes (the SAS 70 Statement on Auditing Standards No. 70).
- access control in public clouds – these apply to verification of users accessing the cloud. The initial registration of a user is a multi-layered procedure, consisting of several overlapping secret questions and answers (e.g. the user's credit card details). Other levels of security verification may include invoice address, call-back verification over the phone (the *out of band* mechanism, based on employing a different channel of communication), login and authorization (the password should be strong), access keys (a good practice here is to provide regular key substitution service), X.509 certificates, paired keys (the latter being the most important element of user verification when working in cloud environment instances)

- network security and protection of data in large clouds. Passing the task on to the experts employed by the cloud service provider seems the best approach, since it may be reasonably assumed that the provider will be faster to respond to a potential intrusion attempt, and that the response will be adequate to the risk at hand. System security in public cloud models is verified at many levels: at the level of the host's operating system, at the level of the virtual machine's operating environment or the host system, at the stateful firewall, and at the level of signed API calls (the cloud application programming interface), with each subsequent level supplementing the capacities of its immediate precursor.
- The role and the responsibilities of the application owner. Cloud users themselves are responsible for security at the level of their host machine accessing the virtual instance. Since the users have full admin control over their accounts, services, and applications, they are responsible for basic security measures, such as the use of strong passwords, safe storage of passwords and private keys, as well as regular key substitution. Data stored in clouds should also be sufficiently protected – for example, by encrypting the resources prior to uploading them to the cloud, to make sure they cannot be read or modified during transmission and storage.

Modern organizations – both the IT customers and IT service providers – should strive to identify and recognize all processes, services and resources considered mission-critical or of key importance from the information security viewpoint. They should also perform a reliable analysis of information risks, and take suitable measures and procedures to minimize the risk over the course of the cooperation with external partners. Thus, irrespective of the security solutions on offer by the service provider, they should employ their own, independent backup procedures with respect to sensitive data – such backup may be of great value if the company decides to withdraw from the contract (in such cases, the provider may refuse access to data stored in their system) or if the provider goes bankrupt.

Companies unwilling to put their trust in external providers, despite numerous obvious benefits offered by cloud computing solutions, can always reach for other models, such as those based on insourcing or the private cloud model.

The insourcing solution is based on internal management of IT services. If need arises, the company may purchase the lacking skills on the market for a limited time, for example by contracting additional personnel for the task. In this model, the organization retains its internal IT personnel and infrastructure, trusting in its ability to free the latent potential of its employees for the purpose of improving its IT services and making them more effective. From the viewpoint of the insourcing model, the internal IT department is formally perceived as a provider of services.

In the case of private cloud solutions, the decision to adopt this particular business model is made on the basis of four fundamental factors: security, accessibility, the size of user population, and the effect of scale (Table 1).

Private clouds offer better control and assurance that the resources will not be used by other customers, since they are not shared in public space. However, as any other solution, the private cloud model has its own limitations, such as (own research, based on [13, pp. 119–120]):

Table 1. Premises for adopting a private cloud solution

Factor	Description
Security	Applications require direct control and data protection, for confidentiality and safety reasons (for instance, governmental agencies use dedicated applications for processing of confidential and classified data – it is essential that they be kept from unauthorized access).
Accessibility	Applications require access to a predefined set of processing resources, and this type of access cannot be securely provided in a shared environment.
User population	The organization caters for many users, often in geographically remote locations, and they all require unrestricted access to computational processing resources (private clouds are used, for example, in large telecommunication corporation).
The effects of scale	Data centers and infrastructural resources are readily available, or can be expanded at minimum cost.

Source: own research based on ([10], p. 116).

- limited scale of operation, compared to public clouds,
- the problem of adopting old applications to the cloud structure requirements without redesigning the very architecture of the system,
- limited potential for optimization and innovation of the methods and elements of the system,
- larger operational outlay compared to the public cloud solutions.

Even if the organization does not anticipate any integration with external providers when choosing their outsourcing solution, it may be advisable to keep an open stance in this respect, so that it may smoothly transition to another model if need arises, and not be too restricted with their choice of a potential provider.

7 Conclusions

New technologies, and the resulting new models and instruments for business, generate new and previously unforeseen risks and threats. Changes in company operating environment, brought about by globalization, increased competition, automation and – most of all – computerization, informatization and virtualization – require a new approach to knowledge management and information security in modern organizations.

The paper emphasizes the role of knowledge management in IT outsourcing projects. The analysis of organizational resources of contracting partners, informational needs and processes of information and knowledge management illustrates the vast number of factors that should be taken into account to minimize the risk of improper 'utilization' of knowledge by any of the contracting parties and to provide effective cooperation between partners of the IT outsourcing project.

The most important conclusions that can be drawn from the above considerations include the following (based on [13]):

(1) Organizations deciding on outsourcing its IT area should emphasize proper preparation steps, such as detailed analysis of needs and potential benefits as well as potential risks that may result from this type of organizational change.

(2) Modeling of knowledge management processes in outsourcing relations and formulation of process maps (both for IT processes and information/communication processes) may effectively:

- limit the extent of communication problems with external provider,
- facilitate and streamline the implementation of IT outsourcing,
- improve security of organizational knowledge (through proper recognition and protection of core 'knowledge carriers', protected access to 'results' of the service provision after contract termination, etc.),
- improve information exchange between the organization and the service provider,
- improve information security (especially in cloud computing model),
- facilitate development of organizational knowledge in the course of the outsourcing contract, both in its organizational (project management knowledge) and technological aspect (of particular importance if the organization plans to restore in-house servicing of the outsourced tasks after contract termination),
- improve the outsourcing process in relation to the overall strategic objectives of the organization.

(3) Building the atmosphere of trust and developing conditions for sharing knowledge between the contracting partners boosts the potential for developing sealed knowledge and innovation in cooperation with external companies.

(4) Knowledge management in organizational relations with the IT service providers should be regarded as a module of a wider, comprehensive model of knowledge management formulated with the main purpose of identifying methods, instruments and relations for organizational survival and development, while at the same time limiting its dependence on external parties.

As discussed in this paper, new IT sourcing models, especially cloud computing, offer some opportunities, but even if organizations themselves feel "cloud ready," they must anticipate the capacity requirements in the cloud. They must also be aware of new risks, and manage their IT security in accordance with the new operating conditions. The most important risk areas with respect to modern IT sourcing solutions (similarly to those observed in classical outsourcing models) include: the loss of control over the IT environment, inadequate protection of data, overdependence on external suppliers, the loss of potential to switch back to previous (self-contained) IT services, etc. A decision to adopt a particular IT sourcing solution should be based on such factors as: the size of the organization, the scale of operation, risk propensity, knowledge management model, the adopted information security policy, the personnel strategy, and the budget.

It seems that migration to a cloud model is a good solution for companies intent on maximizing their profits (cloud computing services are decidedly more cost-effective) while at the same time retaining their high standards of security. What makes the cloud computing particularly attractive for business entities is the fact that they can pass most of the IT system security responsibilities on to the service provider. The providers of cloud computing services, being well aware of the fact that security concerns are the most important factor to restrain companies from choosing the cloud model, make huge investments in security solutions and infrastructure, as a way to emphasize their responsible approach to the security of their clients' resources. Companies which – for a number of reasons – are unable or unwilling to rely on external partners with their data and knowledge, can reach for other sourcing models, such as the private cloud model or the insourcing model, to improve their IT effectiveness and both protect and develop internal knowledge in the IT area.

References

1. Oshri, I., Kotlarski, J., Willcocks, L.P.: The handbook of global outsourcing and offshoring, 2nd edn. Palgrave Macmillan Ltd., Basingstoke (2011)
2. Morgan, J.L., Bravard, R.: Inteligentny outsourcing. Sztuka skutecznej współpracy (in Polish), MT Biznes Sp. z o.o., Polska (2010)
3. Liderman, K.: Bezpieczeństwo informacyjne (in Polish), Wydawnictwo Naukowe PWN, Warszawa (2012)
4. Rot, A., Sobińska, M.: IT security threats in cloud computing sourcing model. In: Proceedings of the Federal Conference on Computer Science and Information Systems (2013 Federated Conference on Computer Science and Information Systems (FedCSIS)), pp. 1299–1303 (2013)
5. Sobińska, M.: IT management business model - sourcing IT services. In: K. Perechuda (ed.): Advanced Business Models – in publishing (2014)
6. Sparrow, E.: Successful IT Outourcing. Springer, London (2003)
7. Willcocks, L.P., Lacity, M.C.: The new IT outsourcing landscape From innovation to cloud services. Palgrave Macmillan Ltd., Basingstoke (2012)
8. Firmy lekceważą cyfrowe ataki (in Polish), Puls Biznesu, 27 November 2013 (2013)
9. Viega, J.: Mity bezpieczeństwa IT. Czy na pewno nie masz się czego bać? (in Polish), Helion (2010)
10. Mateos, A., Rosenberg, J.: Chmura obliczeniowa Rozwiązania dla biznesu (in Polish), Helion, Gliwice (2011)
11. Perechuda, K., Sobińska, M.: Information security in IT global sourcing models. In: Proceedings of the Federated Conference on Computer Science and Information Systems (FedCSIS 2014), pp. 1441–1447 (2014). https://fedcsis.org/proceedings/2014/
12. Willcocks, L.P., Sobińska, M.: IT sourcing management in Poland – trends and performance. In: the paper on the 9th Global Sourcing Workshop (February 18–21, 2015 La Thuile, Italy) - in review (2015)
13. Sobińska, M.: Modeling of knowledge management processes in IT outsourcing projects, Informatyka Ekonomiczna. Business Informatics 20/2011, ISSN 1507-3858, red. A. Bąk, A. Rot Wydawnictwo UE we Wrocławiu, Wrocław (2011)
14. Ciesielska, D.: Offshoring usług. Wpływ na rozwój przedsiębiorstwa (in Polish), Wolters Kluwer Polska, Warszawa (2009)
15. Willcocks, L.P., Lacity, M.C.: The Practice of Outsourcing From Information systems to BPO and Offshoring. Palgrave Macmillan Ltd, Basingstoke (2009)
16. Perechuda, K., Sobińska, M.: Zarządzanie informacją i wiedzą w outsourcingu IT (in Polish). In: Korczak, J., Chomiak-Orsa, I., Sroka, H. (eds.) Systemy informacyjne w zarządzaniu przedsiębiorstwem (in Polish) Wydawnictwo Uniwersytetu Ekonomicznego we Wrocławiu, Wrocław (2010)

How Should Digital Humanities Pioneers Manage Their Data Privacy Challenges?

Francis Rousseaux[1(✉)] and Pierre Saurel[2]

[1] Institut de Recherche et de Coordination Acoustique Musique, Paris, France
francis.rousseaux@ircam.fr
[2] Paris-Sorbonne University, Paris, France
pierre.saurel@paris-sorbonne.fr

Abstract. Since Digital Humanities researchers and developers are regularly creating somehow industrial applications concerning international business, it is time for those communities to be aware and make the most of legacy constraints and opportunities.

For instance, let us consider the Computer Music state of the art, and particularly the Music Information Retrieval community and the wonderful algorithms it produces around authorship attribution and style recognition: even if some music style or authorship is finally attributed to some persons, this attribution may not result from a set of computable data somewhere reportable, the information being typically learned (in the sense of Machine Learning, more or less supervised) from dislocated data throughout the big data or the global database, and disseminated in the global programming system. Is this legal?

In Europe and worldwide, Privacy by Design (PbD) is the actual response to protect the fundamental right to data protection and to guarantee the free movement of personal data between business stakeholders or Member States.

Keywords: Digital humanities · Computer music · Music information retrieval · Machine learning · Knowledge management · Right management · Privacy by design · Big data · Authorship attribution · Style recognition

1 Introduction

Since the Digital Humanities researchers and developers are often about to create somehow industrial applications potentially concerning international business [12, 28, 42], it is time for those communities to be aware and make the most of legacy constraints and opportunities.

For instance, let us consider the Computer Music state of the art, and particularly the Music Information Retrieval (MIR) community, as it is typically structured and organized by the International Society of Music Information Retrieval (ISMIR, see http://www.ismir.net/).

ISMIR is now fifteen years old, and getting out of adolescence. After a fast-growing childhood, bottle-fed by the best IT algorithms and the most vitamin-rich signal analysis methods, the International Society for Music Information Retrieval is now addressing a wide range of scientific, technical and social challenges, dealing with processing,

© IFIP International Federation for Information Processing 2015
E. Mercier-Laurent et al. (Eds.): AI4KM 2014, IFIP AICT 469, pp. 75–91, 2015.
DOI: 10.1007/978-3-319-28868-0_5

searching, organizing and accessing music-related data and digital sounds through many aspects, considering real scale use-cases and designing innovative applications, over-flowing its academic-only initiatory aims.

As the emerging MIR scientific community reaches its disciplinary maturity and leads to potential industrial applications of interest to the international business (start-up, Majors, content providers, download or exchange platforms) and to large scale experimentations involving many users in living labs (for MIR teaching, for multicultural emotion comparisons, or for MIR user requirement purposes) the identification of legal issues becomes essential or even strategic.

Among legal issues, those related to copyright and Intellectual Property have already been identified and expressed into Digital Right Management subsystems by the MIR community [8, 27, 33], when those related to security, business models and right to access have been understood and expressed by Information Access [17, 35]. If those domains remain islands beside the MIR continent, Privacy, as another important part of legal issues, is not even a living island in the actual MIR archipelago.

However, Privacy and personal data issues are currently addressed by many Information Technology (IT) communities, aware of powerful and efficient paradigms like Privacy by Design.

2   Privacy by Design: New Challenges in Big Data

Privacy by Design (PbD) was developed by Ontario's Information and Privacy Commissioner Dr. Ann Cavoukian in the 1990s, at the very birth of the future big data phenomenon. This made-in-Ontario solution has gained widespread international recognition, and was recently recognized as a global privacy standard.

2.1   What Has Changed Within the Big Data?

The first radical change is obviously the web. Everyone produces data and personal data. However, the user is not always aware that he provides personal data allowing his identification. For instance, as described by [40], when a user tags or rates musical items, he gives personal information about himself. If a music recommender exploits this kind of user data without integrating strong privacy concepts, he faces legal issues and strong discontent from the users.

The volume of data has been increasing faster than the "Moore's law". This evolution is known as the concept of "Big Data". New data are generally unstructured and traditional database systems such as Relational Database Management Systems cannot handle the volume of data produced by users and by machines & sensors. This challenge was the main driver for Google to define a new technology: the Apache Hadoop File System. Within this framework, data and computational activities are distributed on a very large number of servers. Data are not loaded for being computed, and the result stored. Here, the algorithm is close to the data.

Databases of personal data are no more clearly identified. We can view the situation as combining five aspects:

Explosion of Data Sources. The number of databases for retrieving information is growing dramatically. Applications are also data sources. *Spotify* for instance, embedded in Facebook, provides a live flow of music consumption information from millions of users. Data from billions of sensors will soon be added. This profusion of data does not mean quality. Accessible does not mean legal or acceptable for a user. Those considerations are essential to build reliable and sustainable systems.

Crossing & Reconciling Data. Data sources are no longer islands. Once the user can be identified (cookie, email, customer id), it is possible to match, aggregate and remix data that were previously technically isolated.

Time Dimension. The web has generally a good memory that humans are not familiar with. Data can be public one day and be considered as very private 3 years later. Many users forget they posted a picture after a student party. And the picture has the bad idea to crop up again when you apply for a job. And it is not only a question of human memory: Minute traces collected one day can be exploited later and provide real information.

Permanent Changes. The general instability of the data sources, technical formats and flows, applications and use is another strong characteristic of the situation. The impact on personal data is very likely. If the architecture of the systems changes a lot and frequently, the social norms also change. Users today publicly share information they would have considered totally private few years earlier. And the opposite could be the case.

User Understandability and Control. Because of the complexity of changing systems and complex interactions users will less and less be able to control their information. This lack of control is caused by the characteristics of the systems and by the mistakes and the misunderstanding of human users. The affair of the private Facebook messages appearing suddenly on timeline (Sept. 2012) is significant. Facebook indicates that there was no bug. Those messages were old wall posts that are now more visible with the new interface. This is a combination of bad user understanding and fast moving systems.

Changes in the Information Technology lead to a shift in the approach of data management: from computational to data exploration. The main question is "What to look for?" Many companies build new tools to "make the data speak" and usually find personal data. This is the case considering the underlying trend of heavily personalized marketing. Engineers using the big data usually deal with existing personal data and build systems that produce new personal dataflow.

2.2 Foundations of Privacy by Design

According to its inventor Ann Cavoukian[1], "Privacy by Design is an approach to protect privacy by embedding it into the design specifications of technologies, business practices, and physical infrastructures. That means building in privacy up front – right into

---

[1] http://www.ipc.on.ca/images/Resources/7foundationalprinciples.pdf.

the design specifications and architecture of new systems and processes. PbD is predicated on the idea that, at the outset, technology is inherently neutral. As much as it can be used to chip away at privacy, it can also be enlisted to protect privacy. The same is true of processes and physical infrastructure".

1. Proactive not Reactive; Preventative not Remedial. The PbD approach is characterized by proactive rather than reactive measures. It anticipates and prevents privacy invasive events before they happen. PbD does not wait for privacy risks to materialize, nor does it offer remedies for resolving privacy infractions once they have occurred — it aims to prevent them from occurring. In short, PbD comes before-the-fact, not after.
2. Privacy as the Default Setting. We can all be certain of one thing — the default rules! PbD seeks to deliver the maximum degree of privacy by ensuring that personal data are automatically protected in any given IT system or business practice. If an individual does nothing, their privacy still remains intact. No action is required on the part of the individual to protect their privacy — it is built into the system, by default.
3. Privacy Embedded into Design. PbD is embedded into the architecture of IT systems and business practices. It is not bolted on as an add-on, after the fact. The result is that privacy becomes an essential component of the core functionality being delivered. Privacy is integral to the system, without diminishing functionality.
4. Full Functionality — Positive-Sum, not Zero-Sum. PbD seeks to accommodate all legitimate interests and objectives in a positive-sum "win-win" manner, not through a dated, zero-sum approach, where unnecessary trade-offs are made. PbD avoids the pretense of false dichotomies, such as privacy vs. security, demonstrating that it is possible to have both.
5. End-to-End Security — Full Lifecycle Protection. PbD, having been embedded into the system prior to the first element of information being collected, extends securely throughout the entire lifecycle of the data involved — strong security measures are essential to privacy, from start to finish. This ensures that all data are securely retained, and then securely destroyed at the end of the process, in a timely fashion. Thus, PbD ensures cradle to grave, secure lifecycle management of information, end-to-end.
6. Visibility and Transparency — Keep it Open. PbD seeks to assure all stakeholders that whatever the business practice or technology involved, it is in fact, operating according to the stated promises and objectives, subject to independent verification. Its component parts and operations remain visible and transparent, to users and providers alike. Remember, trust but verify.
7. Respect for User Privacy — Keep it User-Centric. Above all, PbD requires architects and operators to keep the interests of the individual uppermost by offering such measures as strong privacy defaults, appropriate notice, and empowering user-friendly options. Keep it user-centric.

2.3   Prospects for Privacy by Design

In Europe and worldwide [23], Privacy by Design is considered as the best current operational response to both protect the fundamental right to data protection and guarantee the free flow of personal data between business stakeholders or Member States.

Thus, at the time of digital data massive exchange through networks, privacy by design is a key-concept in legacy [32, 36, 43, 46].

For instance in Europe, where this domain has been directly inspired by the Canadian experience, the European Community[2] affirms that "Privacy by Design means that privacy and data protection are embedded throughout the entire life cycle of technologies, from the early design stage to their deployment, use and ultimate disposal".

Privacy by Design becomes a reference for designing new systems and processing involving personal data. It becomes even an essential tool and constraint for these designs whereby it includes signal analysis methods as long as these analyses integrate or produce personal data.

Concerning the scientific community, we can recall two main points:

Processing relative to historical, statistical and *scientific research* purposes, falls under specific conditions defined by article 83 of the "Safeguarding Privacy in a Connected World" European law that facilitates the use of personal data in certain cases. This article defines two specific exceptions, i.e. when: (i) these processing cannot be fulfilled otherwise and (ii) data permitting the identification are kept separately from the other information, or when the bodies conducting these data respect three conditions: (i) consent of the data subject, (ii) publication of personal data is necessary and (iii) data are made public;

Penalties in case of non-compliance are severe. As long as processing is not compliant, these penalties are the same whether the algorithms and the processing used in real business are issued from the research community or not. The supervisory authority "shall impose a fine up to €1,000,000 or, in case of a company up to 2 % of its annual worldwide turnover".

2.4 Europe vs. United States: Two Legal Approaches

Europe regulates data protection through one of the highest State Regulations in the world [16, 31] when the United States lets contractors organize data protection through agreements supported by consideration and entered into voluntarily by the parties. These two approaches are deeply divergent. United States lets companies specify their own rules with their consumers while Europe enforces a unique regulated framework on all companies providing services to European citizens. For instance any company in the United States can define how long they keep the personal data, when the regulations in Europe would specify a maximum length of time the personal data is to be stored. And this applies to any company offering the same service.

A prohibition is at the heart of the European Commission's Directive on Data Protection (95/46/CE – The Directive) [16]. The transfer of personal data to non-European Union countries that do not meet the European Union adequacy standard for

---

2 "Safeguarding Privacy in a Connected World – A European Data Protection Framework for the 21st Century" COM (2012) 9 final.

privacy protection is strictly forbidden [16, article 25][3]. The divergent legal approaches and this prohibition alone would outlaw the proposal by American companies of many of their IT services to European citizens. In response the U.S. Department of Commerce and the European Commission developed the Safe Harbor Framework (SHF) [23, 41]. Any non-European organization is free to self-certify with the SHF and join.

A new Proposal for a Regulation on the protection of individuals with regard to the processing of personal data has been adopted on 12th March 2014 by the European Parliament [31]. The Directive allows adjustments from one European country to another and therefore diversity of implementation in Europe when the regulation is directly enforceable and should therefore be implemented directly and in the same way in all countries of the European Union. This regulation should apply in 2016. This regulation enhances data protection and sanctions to anyone who does not comply with the obligations laid down in the Regulation. For instance [31, article 79] the supervisory authority will impose, as a possible sanction, a fine of up to one hundred million Euros or up to 5 % of the annual worldwide turnover in case of an enterprise.

Until French law applied the 95/46/CE European Directive, personal data was only defined considering sets of data containing the name of a natural person. This definition has been extended; the 95/46/CE European Directive (ED) defines 'personal data' [16, article 2] as: "any information relating to an identified or identifiable natural person ('data subject'); an identifiable person is one who can be identified, directly or indirectly, in particular by reference to an identification number or to one or more factors specific to his physical, physiological, mental, economic, cultural or social identity".

For instance the identification of an author through the structure of his style as depending on his mental, cultural or social identity is a process that must comply with the European data privacy principles.

3   The Way ISMIR Supports Legal Issues

Let us look ahead ISMIR works from the point of view of pro-activity about data and especially about the legal and personal data.

3.1   ISMIR Works Regarding Privacy Issues

Is it possible to get a clear view of ISMIR evolution regarding the legal themes and especially privacy from the year 2000 to the year 2013 without going into technical details? Let us begin by some 2000 to 2013 sessions raw comparisons of 1 Topics mentioned within the official calls for contributions and papers, 2 Topics that characterize the accepted papers, posters and interventions.

---

[3] Argentina, Australia, Canada, State of Israel, New Zealand, United States – Transfer of Air Passenger Name Record (PNR) Data, United States – Safe Harbor, Eastern Republic of Uruguay are, to date, the only non-European third countries ensuring an adequate level of protection: http://ec.europa.eu/justice/data-protection/document/international-transfers/adequacy/index_en.htm

In 2000, one category out of the eleven offered, concerned legal aspects: "Intellectual property rights issues". And in the forthcoming edition there is one category "social, legal, ethical and business issues" out of a total of thirty-two. Looking simply at those figures, we can note that personal data and rights issues occupy a secondary place on ISMIR's agenda. But, as we see below, those questions are very transversal to many research topics.

Let us now have a look at the topics in the ISMIR titles of accepted papers, posters and interventions (Table 1).

Table 1. A word-cloud fed with ISMIR 2000 & 2001 scientific papers

At first glance, there is no significant change in the scope of legal, right or personal data: only tree interventions mention the term 'legal' or 'right' in their title, no publication mentions explicitly "Personal Data" or "Privacy". However, behind many themes of the different ISMIR works, personal data are involved in one way or another. The most obvious one concerns the "Personal music libraries" and "recommendation": more than 30 papers and posters deal with those topics as being their main topic. How to recommend music to a user or analyze their personal library without tackling privacy? And how to work on "Classification" or "Learning", producing 130 publications without considering users throughout their tastes and their style? (Table 2).

Table 2. The corresponding 2011 & 2012 word-cloud

3.2   Why the Lack of Pro-Activity Regarding Legal Issues Can Lead to Failure?

We may agree that the MIR scientific community, as noticeable through ISMIR successive publications, is deepening its scientific objects and sub-domains, creating powerful algorithms, features and metadata, considering research as its main activity.

But business is not far away, and will completely reorganize the traditional stakeholders' models, requiring user involvement to design recommendation systems and to extract knowledge and evaluation. For having ignored the necessity to address legal issues, and particularly privacy issues, other IT innovative areas have already collapsed.

For instance, the Digital Right Management for digital music was an attempt from the producers to recover their intellectual property on music already largely shared by users on the web. It is clear that the legal aspects were not integrated "by design" by the different stakeholders in the context of music stored in digital lossless files, exchangeable worldwide, easy to be copied.

In the context of personal data, the principle of "End to security" is regularly in the limelight. For instance in 2011 millions of PlayStation accounts where hacked. The consequences for Sony have been considerable and, two years later, prosecutions are still in progress. In Great Britain, Sony was condemned in 2013, and during the trial, the authorities mentioned that: "polls conducted after the breach suggested a greater awareness of the risks in handing over personal data". Once again, the effort here is conducted after the issue.

Those two simple examples only show the beginnings of the problems that will arise if private data management is not proactive. The new capabilities of the Information Technology announce a much more complicated world in terms of personal data.

3.3   Actual MIR Practices are PbD-Compatible but not PbD-Compliant

Just as music/sound-based system design is not the MIR core target, constraints related to architecture design (technical constraints or related to user interfaces) are not in the core focus of MIR researchers either. That is why, even though this notion is more than twenty years old, Privacy by Design — as an intersection between IT content and method — has not directly involved the ISMIR contributors yet, no more than the International Computer Music Conference ones.

Furthermore, most of ISMIR contributions are still research oriented, in the sense of Article 83 of the "Safeguarding Privacy in a Connected World". To say more about that intersection, we need to enter into a cross survey of the ISMIR scientific ten-years production, throughout the main PbD Foundational Principles (FP).

FP6 (transparency) and FP7 (user-centric) are most of the time fully respected among the MIR community as source code and processing are often (i) delivered under GNU like licensing allowing audit and traceability (ii) user-friendly.

However, as long as PbD is not embedded into Design, FP3 cannot be fulfilled and accordingly FP2 (default setting), FP5 (end-to-end), FP4 (full functionality) and FP1 (proactive) cannot be fulfilled either. Without any PbD embedded into Design, there are no default settings (FP2), you can not follow and end-to-end approach (FP5), you can not define full functionality regarding to personal data (FP4) and you can even less be proactive. Principle of pro-activity (FP1) is the key principle. If you fulfill FP1 you can define the default settings (FP2), be full functional (FP4) and define an end-to-end process (FP5).

Actual MIR Practices claim to be relatively neutral to data privacy and are compatible with PbD. The MIR Practices could be compliant to PbD as long as they would fulfill the FP1 principle of pro-activity.

4 How ISMIR Builds New Kind of Personal Data

ISMIR methods apply algorithms to data. Most of the time these non linear methods use inputs to build new data which are outputs or data stored inside the algorithm, as weights for instance in a neural net.

The Gamelan Project is a study case where machines produce new data and new personal data from inputs.

4.1 A Case Study: The Gamelan Project

Gamelan (see http://projet-gamelan.fr/) was an Industrial Research category research project coordinated by IRCAM, gathering INA, EMI Music France and UTC, and supported by the French *Agence Nationale pour la Recherche*. The project began in November 2009 and lasted 48 months.

Digital studios involve a great amount of traceable processes and objects, because of the intense producer-device interactions during contents production. The important flow of these traces called for a system to support their interpretation and understanding. The Gamelan research teams has studied and developed such a system in the digital music production context, towards musical object and process reconstitution. We were aiming at combining trace engineering, knowledge modeling and knowledge engineering, based on the differential elaboration of an ontology, standard formats and common knowledge management tools.

Involving professional users, the Gamelan research teams succeeded in applying this system to several different real use cases, put forward by different kind of end users, and we are now able to discuss some hypothesis about trace-based knowledge management, digital music preservation and reconstitution, opening on to some considerations about artistic style, and to the specification of the next generation prototypes that music industry would need to develop.

Gamelan is also the name of the developed software environment, built upon the production ecosystem, to address the reconstitution issue of digital music production, by combining trace engineering, knowledge modelling and knowledge engineering. Most of the time reconstitution is relegated afterwards. Gamelan aims at reconstructing the composer-system interactions that have led to the creation of a work of art that is about to leave the production studio. The purposes of reconstruction concern long-term preservation, repurposing, versioning and evolution of the work of art, and more generally the disclosure of the contingencies of its initial outcome.

In the music production studio, everything is about creativity [15, 19, 45]. Until now, music tools design has mainly focused on the making of the final product, because the very first aim of the studio is to provide the creator with efficient means to make and shape the musical object he or she came in the studio for. But this requisite priority on creativity has overshadowed another need that appears later: reconstitution.

Of course, creativity empowering raises tough challenges to work out. For instance, on the conceptual side, bridging the gap between creative thinking and application interfaces remains a challenging issue [2, 11, 34], while on the technical side, the heterogeneity of tools, systems, components, protocols and interfaces keeps renewing difficulties for the management of production environment [13].

A creator finishing his or her work in a studio marks the end of the production process: the so-awaited object is finally there, thus the creator, the producer, the sound engineer and all the people involved are happy or at least relieved; the goal is reached and the story reaches its end. However, at this very moment, because the final object is there, no one wonders about its reconstitution.

But — say ten years later — when "back-catalog" teams of music companies want to edit some easy to sell Greatest Hits at up-to-date audio formats, mining the musical archives is no longer easy. Back to the reachable-recorded digital files, it may be painful to figure out which one of the bunch of files left is the one needed. File dates and file names are not trustable (Fig. 1).

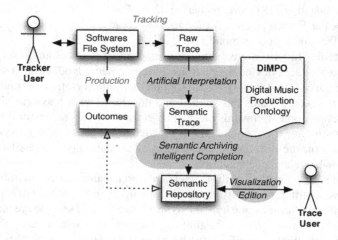

Fig. 1. Gamelan: the interaction graph

Closer in time — say two months after the production — the simple task of collecting vital information on the contributors who actually worked on the project may turn into a real problem. A musician may replace another without logging his/her name. Or a name is missing because we only have the nickname and we don't have the phone number either. There is a whole set of information on contributions (name, role, time spent, etc.) necessary to manage salaries, rights and royalties that regularly proves hard to collect afterwards. Evidently, this kind of information would be far easier to collect directly at production time.

On the surface, nothing to do with privacy and personal data! But in fact, and it is typically the case as soon as a complex person-software device is involved, this type of project invites us to rethink classical approaches and qualifications of privacy issues.

4.2   Collecting Traces to Build a Knowledge Model in the Gamelan Project

The Gamelan project exemplifies several of the many R&D emerging questions that are raised in the digital audio processing domain.

First of all reconstitution requires to collect traces during the production process itself. Automatically-collected software traces differ from human-entered traces. The former can be seamlessly collected through automatic watching components, with interfaces traces and logs as heuristic material, while the latter inevitably requests a human contributor for information that cannot be automatically captured or inferred from automatic traces. A full-production tracking environment would resemble Living Labs, towards a Living Studio.

Secondly, these traces call for an appropriate knowledge model. To stay as little invasive as possible, such a model should provide means to determine which information is worth to ask humans during the production or not compared to the creativity disturbing cost. Without a knowledge model, it would not be possible to interpret the traces or to determine the kind of traces worth capturing. To achieve this model, professional knowledge must be identified, listed and characterized with experts, defining a digital music production Knowledge Level (Fig. 2).

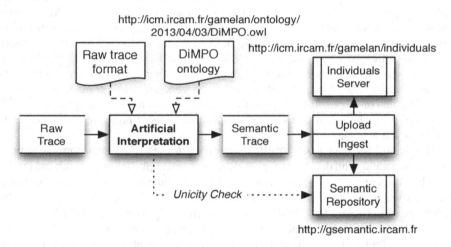

Fig. 2. Gamelan: the different functionalities

Within Gamelan, traces from used operating system and from used professional applications are extracted, semantic networks dealing with typical digital audio composition acts are involved towards some abstraction of those traces, but personal data are nowhere considered: some real time digital audio flow are involved, transformed on the fly by creative acts that sign the composer particular style and their artistic singularity.

The composer style, as part of built up personal data, often not even named, is computed to support the Gamelan reconstruction process: what is interesting to reconstitute has something to do with the abstract truth of the given piece of art and its stylistic genesis. To understand that "the composer is currently testing a sample within the whole piece framework" is more efficient than being aware of a succession of cut-paste-listen actions that has to be generalized.

Thus some personal data, like artistic style [1, 7, 13, 18] are built up on the fly, relatively to processing algorithms, knowledge bases and title repositories [9], evolving from the system experience itself, and only known by the system. The ultimate target is clearly the style-recognition [22] of the creator, viewed as the correlation between their practice and the character of their work of art.

4.3 Information Retrieval Produces Personal Data from Inputs

Most of the time the ISMIR developed methods create new outputs that are personal data or include personal data. These data may be used through streaming without being stored. These data could be so evanescent that they could not be reproduced at all, depending on the specific situation of the performance processing and of the music listener. These data are personal data as they depend on the specific listener or the specific musician.

Authorship attribution creates new personal data, whatever the algorithmic way you build this attribution.

Some of these non-linear algorithms encode personal data implicitly. For instance Kohonen maps, Neural Nets, Hidden Markov Models (HMM), Bayesian Maps create and store data [21], which are the weights of the network (especially the weights of the hidden layer) or a set of probability [20]. In that case these data are personal data integrated inside the processing and encoding for instance the author attribution. Moreover this technical way to encode these new personal data converges thanks to the big data.

Fast Fourier Transform, Wavelets, and more generally algorithms that consist in projecting data on a specific basis produce personal data in the same way. Results of the projection are new data, which are personal as long as these projections depend on the listener or the musician.

This is still the case when a MIR research team uses some algorithms to support heuristic reasoning and decision support, dealing for example with authorship attribution or style recognition. In this case, music could be considered as a sequence that can be randomly walked or, better, turned into some semi-predictive program solution by taking advantage of (let's say) local HMM amnesia to (let's say) converge towards a relevant future. Then a musical sequence can be considered as a solution to some Constraints Satisfaction Problem [6, 30], and no one is henceforth able to separate data from processing, a fortiori personal data from processing.

5 MIR Communities: An Emblem of Digital Humanities Ones?

5.1 Style Identification Algorithms in Actual MIR: A Classification

Among the algorithms that identify artists creating a style of their own, we can distinguish two meta-heuristics groups. The first one is based on interconnections of databases and more generally on the big data. The second one is based solely on the intrinsic properties of music data.

In this section we propose a classification of algorithms and heuristics allowing the identification of the creators of musical data exclusively on the basis of this second approach.

It is usual to consider the style of an artist is available at three levels: patterns, meanings and feelings (cf. [3], p. xii). For each of these levels, the algorithms implemented enter the structured list given below which can be used as read gate to analyze style recognition algorithms.

Symbolic Approaches

Shape Grammars, Rules Production and Combination. *Shape Grammars, Rules Production and Combination.* Shape grammars are examples of symbolic approach to style production (Mitchell 1990). As a reverse problem it deals with style identification. Shape grammars were first introduced [37] for painting and sculpture. Later on it has been used for producing new designs in architecture [38]. Even useable pieces of software were developed in this field [29].

Shape grammars are a kind of results obtained by rules production and combination. In this same class of algorithms we can put L-systems [24].

As reverse problem let's consider the results of the rules production system and let's try to find the kind of rules that could produce the same shapes. Find theses rules and you find the style of the production system.

Combination of Patterns. Another way to deal within this symbolic approach is to combine rules but defined short patterns. This is the kind of algorithms produced by David Cope with the so-called EMI project [14].

Optimization Methods. Some of the algorithms used to categorize features are dealing with optimization techniques. Among those we find Support Vector Machines [42] that is used do define a *hyper plane* separating two sets of examples (positive and negatives ones). This *hyper plane* is the set of points that do maximize the distance between the two sets of examples.

Yang and Pedersen [45] proposed a large comparison between those technical optimization methods.

Algorithms Based on Topology. Some other algorithms use topology and connectivity to categorize features and style. Among those we can consider Kohonen cards so-called Self Organizing Maps [26].

For instance Jupp and Gero [24] do apply these algorithms to categorize styles.

Non Deterministic Algorithms

Frequency and Statistical Methods. One of the most intuitive technique to define an artist's style is to build a dictionary of the specific « atoms » of creation he uses. These atoms can be words, music notes, etc. The algorithm consists on building this dictionary and storing the specific personal frequencies of uses of the atoms by the artist.

Argamon and Koppel [4] solve the authorship verification problem by using this key algorithm.

Karlgren [25] explains how these algorithms, defined to detect sylitic variations based on different low-level features can not directly be used for variations of a higher level of abstraction [10].

Markov Models and Especially Hidden Markov Models. In the simplest case, Markov Models consist on a set of states and transition probabilities from one state to the other ([3], Chaps 7 & 10). More sophisticated Markov Models, also called Hidden Markov Models (HMM) consist on Markov Models with hidden states. Hidden states are kind of internal states that cannot be directly perceived.

These models were widely used to model and categorize styles. For instance they were used for instance to identify and distinguish Beethoven sonatas and Mozart sonatas [2].

Assayag et al. [5] apply these techniques to Anticipatory Learning, which mixes these HMM and Q-learning [39].

5.2   From MIR R&D Communities to Digital Humanities Ones

In the current legal frameworks, Personal Data are still considered as a particular kind of data, as opposed to processing according to the classical Information Technology paradigm. By seriously considering the PbD methods and technologies for mastering and appropriation, the MIR community will probably rediscover that the paradigmatic data/processing separation has been finally overcome, as soon as many MIR algorithms raise their results on the fly from digital music flows, often in real-time: thus the MIR community will naturally join and lead the conducting of the PbD contemporary concept towards more advanced concepts, able to take note that sometimes, personal data can be outputs from some authorship attribution artificial system, made of complex person-machine interactions and act accordingly.

The time where data (on the one hand) and processing (on the other hand) were functionally independent, formally and semantically separated, has ended. Nowadays, MIR researchers currently use algorithms that support effective decision, supervised or not, without introducing 'pure' data or 'pure' processing, but building up acceptable solutions together with heuristic knowledge that cannot be reduced to data or processing.

This means that new research fields do not separate data and processing anymore. This can be done in different ways. In many circumstances, the MIR community develops new personal data using the whole range of data analysis and data building algorithms. The MIR community is especially well positioned to identify the new personal data produced through these algorithms.

From this respect, MIR is a good emblem of what is currently happening within the whole Digital Humanities R&D communities.

6   Conclusion

6.1   When Some Process Lead to Direct or Indirect Personal Data Identification

Methodological Recommendations. Digital Humanities researchers and developers could first audit their algorithm and data, and check if they are able to identify a natural person (two first sets of our classification). If so they could use the Safe Harbor Framework which could already be an industrial challenge for instance regarding Cyber Security (P5). Using the Privacy by Design methodology certainly leads to operational solutions in these situations.

6.2   When Some Process May Lead to Indirect Personal Data Identification Through Some Complex Process

In many circumstances, the Digital Humanities researchers and developers community develops new personal data on the fly, using the whole available range of data analysis and data building algorithm. Then researchers could apply the Privacy by Design methodology, to insure that no personal data is lost during the system design.

Here PbD is not a universal solution because the time when data (on the one hand) and processing (on the other hand) were functionally independent, formally and semantically separated, has ended. Nowadays, Digital Humanities researchers and developers currently use algorithms that support effective decision, supervised or not, without introducing 'pure' data or 'pure' processing, but building up acceptable solutions together with machine learning [21] or heuristic knowledge that cannot be reduced to data or processing: The third set of personal data may appear, and raise theoretical scientific problems.

Political Opportunities. The Digital Humanities community has a political role to play in the data privacy domain, by explaining to lawyers — joining expert groups in the US, UE or elsewhere — what we are doing and how we overlap with the tradition in style description, turning it into a computed style genetic, which radically questions the analysis of data privacy traditions, cultures and tools.

Future Scientific Works. In addition to methodological and political ones, we face purely scientific challenges, which constitute our research program for future works. Under what criteria should we, as Digital Humanities practitioners, specify when a set of data allows an easy identification and belongs to the second set or on the contrary is too complex or allows a too uncertain identification so that we would say that these are not personal data? What characterizes a maximal subset from the big data that could not ever be computed by any Turing machine to identify a natural person with any algorithm?

References

1. Ackoff, R.: From data to wisdom. J. Appl. Syst. Anal. **16**(1), 3–9 (1989)
2. Alamkan, C., Birmingham, W.P., Simoni, M.H.: Stylistic Structures: An initial Investigation of the Stochastic Generation of Tonal Music. University of Michigan, Computer Science and Engineering Division, Department of Electrical Engineering and Computer Science (1999)
3. Argamon, S., Burns, K., Dubnov, S. (eds.): The Structure of Style. Springer, Heidelberg (2010)
4. Argamon, S., Koppel, M.: The rest of the story: finding meaning in stylistic variations. In: Argamon, S., Burns, K., Dubnov, S. (eds.) The Structure of Style. Springer, Heidelberg (2010)
5. Assayag, G., Bloch, G., Cont, A., Dubnov, S.: Interaction with machine improvisation. In: Argamon, S., Burns, K., Dubnov, S. (eds.) The Structure of Style. Springer, Heidelberg (2010)
6. Aucouturier, J.-J., Pachet, F., Roy, P., Beurivé, A.: Signal+Context = Better Classification. In: Proceedings of the International Symposium on Music Information Retrieval (2007)

7. Barki, H., Hartwick, J.: Measuring user participation, user involvement, and user attitude. MIS Q. **18**, 59–82 (1994)
8. Barlas, C.: Beating babel - identification, metadata and rights, Invited Talk. In: Proceedings of the International Symposium on Music Information Retrieval (2002)
9. Bertin-Mahieux, T., Ellis, D.P., Whitman, B., Lamere, P.: The million song dataset. In: Proceedings of the International Symposium on Music Information Retrieval (2011)
10. Biber, D.: A typology of English texts. Linguistics **27**, 3–43 (1989)
11. Bonnardel, N., Marmèche, E.: Evocation processes by novice and expert designers: towards stimulating analogical thinking. Creativity Prod. Innov. **13**(3), 176–186 (2004)
12. Bonnardel, N., Zenasni, F.: The impact of technology on creativity in design: an enhancement? Creativity Prod. Innov. **19**(2), 180–191 (2010)
13. Carney, J.D.: The style theory of art. Pac. Philos. Q. **72**(4), 272–289 (1991)
14. Cope, D.: Experiments in music intelligence. In: Proceedings of the International Computer Music Conference, San Francisco (1987)
15. Cope, D.: Computer Models of Musical Creativity. MIT Press, Cambridge (2006)
16. Directive (95/46/EC) of 24 Oct. 1995, Official Journal L281, 23/11/1995, pp. 31–50. http://eur-lex.europa.eu/LexUriServ/LexUriServ.do?uri=CELEX:31995L0046:en:HTML
17. Downie, J.S., Futrelle, J., Tcheng, D.: The international music information retrieval systems evaluation laboratory: governance, access and security. In: Proceedings of the International Symposium on Music Information Retrieval (2004)
18. Dubnov, S., Assayag, G., Lartillot, O., Bejerano, G.: Using machine-learning methods for musical style modeling. Computer **36**(10), 73–80 (2003)
19. Edmonds, E.A., Weakley, A., Candy, L., Fell, M., Knott, R., Pauletto, S.: The studio as laboratory: combining creative practice and digital technology research. Int. J. Hum Comput Stud. **63**(4), 452–481 (2005)
20. Fienberg, S.E.: The relevance or irrelevance of weights for confidentiality and statistical analyses. J. Priv. Confidentiality **1**(2), 183–195 (2009)
21. Gkoulalas-Divanis, A., Saygin, Y., Verykios, V.S.: Special issue on privacy and security issues in data mining and machine learning. Trans. Data Priv. **4**(3), 127–187 (2011)
22. Grachten, M., Widmer, G.: Who is who in the end? recognizing pianists by their final ritardandi. In: Proceedings of the International Symposium on Music Information Retrieval (2009)
23. Greer, D.: Safe harbor - a framework that works. Int. Data Priv. Law **1**(3), 143–148 (2011)
24. Jupp, J., Gero, J.: Let's look at style: visual and spatial representation and reasoning design. In: Argamon, S., Burns, K., Dubnov, S. (eds.) The Structure of Style. Springer, Heidelberg (2010)
25. Kalgren, J.: Textual stylistic variation: choices, genres and individuals. In: Argamon, S., Burns, K., Dubnov, S. (eds.) The Structure of Style. Springer, Heidelberg (2010)
26. Kohonen, T.: Self-Organizing Maps, vol. 30. Springer, Heidelberg (1995)
27. Levering, M.: Intellectual property rights in musical works: overview, digital library issues and related initiatives, Invited Talk. In: Proceedings of the International Symposium on Music Information Retrieval (2000)
28. Lubart, T.: How can computers be partners in the creative process: classification and commentary on the special issue. Int. J. Hum Comput Stud. **63**(4), 365–369 (2005)
29. McKay, A., Chase, S., Shea, K., Chau, H.H.: Spatial grammar implementation: from theory to useable software. Artif. Intell. Eng. Des. Anal. Manuf. (AI EDAM) **26**(02), 143–159 (2012)
30. Pachet, F., Roy, P.: Hit song science is not yet a science. In: Proceedings of the International Symposium on Music Information Retrieval (2008)

31. Proposal for a Regulation on the protection of individuals with regard to the processing of personal data was adopted the 12 March 2014 by the European Parliament. http://www.europarl.europa.eu/sides/getDoc.do?type=TA&reference=P7-TA-2014-0212&language=EN

32. Reiss, J.D., Sandler, M.: Audio issues in MIR evaluation. In: Proceedings of the International Symposium on Music Information Retrieval (2004)

33. Reding, V.: The European data protection framework for the twenty-first century. Int. Data Priv. Law **2**(3), 119–129 (2012)

34. Schön, D.A.: The Reflective Practitioner: How Professionals Think in Action, vol. 5126. Basic books, New York (1983)

35. Seeger, A.: I found it, how can i use it? - dealing with the ethical and legal constraints of information access. In: Proceedings of the International Symposium on Music Information Retrieval (2003)

36. Slavkovic, A.B., Smith, A.: Statistical and learning-theoretic challenges in data privacy. J. Priv. Confidentiality **4**(1), 1–243 (2012). Special Issue

37. Stiny, G., Gips, J.: Shape grammars and the generative specification of painting and sculpture. In: Information Processing, vol. 71, pp. 1460–1465. North-Holland Publishing Company (1972)

38. Stiny, G.: Introduction to shape and shape grammars. Environ. Plan. B Plan. Des. **7**(3), 343–351 (1980)

39. Sutton, R.S., Barto, A.G.: Reinforcement Learning: An Introduction. MIT Press, Cambridge (1998)

40. Symeonidis, P., Ruxanda, M., Nanopoulos, A., Manolopoulos, Y.: Ternary semantic analysis of social tags for personalized music recommendation. In: Proceedings of the International Symposium on Music Information Retrieval (2008)

41. U.S.–EU Safe Harbor. http://www.export.gov/safeharbor/eu/eg_main_018365.asp

42. Vapnik, V.: The Nature of Statistical Learning Theory. Springer, New York (1999)

43. Wright, D., Wadhwa, K.: Introducing a privacy impact assessment policy in the EU member states. Int. Data Priv. Law **3**(1), 13–28 (2012)

44. Wang, F.Y.: Is culture computable? Intell. Syst. IEEE **24**(2), 2–3 (2009)

45. Yang, Y., Pedersen, J.P.: A comparative study on feature selection in text categorization. In: Proceedings of the Fourteenth International Conference on Machine Learning (ICML 1997), pp. 412–420 (1997)

46. Zaslavsky, A.M.: Privacy and the statistician: what do we need to know to certify nondisclosure? J. Priv. Confidentiality **3**(2), 83–90 (2011)

Usability of Knowledge Portals for Exclusives in Local Governments

Krzysztof Hauke[✉], Mieczysław L. Owoc, and Maciej Pondel

Wroclaw University of Economics, Komandorska 118/120, 53-345 Wrocław, Poland
{krzysztof.hauke,mieczyslaw.owoc,maciej.pondel}@ue.wroc.pl

Abstract. Exclusion phenomenon in common understanding denotes processes in which members of society or groups of people are permanently blocked from resources (mostly considered as social exclusion). No doubts, such sort of phenomena is observed as unwanted not only from "outsiders" but also from local and global society points of view. The exclusion (and its antonym inclusion) phenomenon can be investigated including many perspectives: starting from identifying exclusion as a process, through multidimensional aspects up to solutions available in the domain.

In order to be successful in overtaking this phenomenon groups and institutions involved in this process should be supported by ICT solutions. The paper consists of six parts which gradually present context of the problem and proposed solutions. After short introduction concerning research background the discussed concept of exclusion processes and knowledge portals are presented. In the main section a general idea of knowledge portal for exclusives is proposed and specialty of these portal for regional implementation in the Silesia agglomeration is discussed. It creates opportunities for formulation final conclusions about the necessity and usability of the developed platform.

1   Introduction

People living in the era of information society, know more and more about a man, groups and the whole nations, about their needs and levels of satisfaction of the human beings alone or of human associations. If so, natural tendency - present in democracy - to treat all citizens equally becomes very important. Any form of discrimination of some group of people is against the democratic order and societies and governments try to counteract with this unwanted phenomenon.

The roots of the exclusion phenomena "discovering" (or better reflexion on human sense of equality and justice) can be found in the discourse in France in the mid 1970 s (see: N. Rawal review of social inclusion and exclusion - [11]). A bit later H. Silver (see: [1]) formulated three paradigms of social exclusion: *solidarity* (stressing social dimension of human interactions), *specialization* (discovering exclusion as a form of discrimination) and *monopoly* (interpreting exclusion as a consequence of the existing group monopolies). Anyway in older and later approaches to the phenomena research many aspects of social exclusion and inclusion were analysed.

In order to be successful, a problem of exclusion should be investigated, reasons of its occurrence should be discovered and solutions for inclusion should be proposed.

© IFIP International Federation for Information Processing 2015
E. Mercier-Laurent et al. (Eds.): AI4KM 2014, IFIP AICT 469, pp. 92–106, 2015.
DOI: 10.1007/978-3-319-28868-0_6

According to European Union policies the poverty and exclusion problems are very important and responses could be projects prepared in the Europe 2020 strategy (Societal challenges section in Horizon 2020 programme, see: [14]).

Two examples should be stressed as promising solutions in the discussed area:

- GSDRC – Applied Knowledge Services devoted to maintaining knowledge about exclusion phenomena (see: [12])
- Exclusion-Inclusion Suburbs – prepared for knowledge services essential in city environments (see: [13]).

In both cases presented solutions are limited to selected phases or areas of exclusion phenomena. Therefore lack of common platform developed for the whole community seems to be obvious.

The paper is managed as follows. In the next section theoretical background of the investigated phenomena is described including nature of exclusion and inclusion phenomena, reasons of its occurrence is discussed and multidimensional characteristics of investigation is stressed. An essence of knowledge portals developed for modern society is presented in the subsequent section with focus on society needs and functionality of such portals, offered architectures and applications useful for different segments of the society. The most innovative part of the research is presented in the main section of the paper devoted to concepts of the knowledge portal addressed to exclusives covering: assumptions, architecture and examples of supported tasks. Implementation aspects essential for local governments in one of Polish agglomeration are discussed in the next section. The paper ends with conclusion remarks and future research.

2 Exclusion Processes a Research Challenges

No doubts, exclusion as a phenomenon seems to be very important and difficult problems to solve in modern society. There are many contexts in which exclusionary processes can occur including different objects, or time- and territory- aspects. At least two approaches should be taken into account *actors-oriented* and *capability-oriented*.

In accordance to the first one the critical thing is: relationships between entities (e.g. "actors") essential for the exclusion. It is very important in understanding the "exclusion" idea as a concept. According to R Saith: [2] *"Social exclusion is a socially constructed concept, and can depend on an idea of what is considered 'normal'"*. So here the crucial topic of understanding of the discussed phenomenon is a definition of normality which, in turn, depends on standard living, hierarchy of values, assumed criteria of society organization and the like what finally can be identified with "actors".

On the other hand A. Sen (see: [3]) keeps that an essence of social exclusion relates to *'functionings'* and *'capabilities'* concepts. *Functionings* denotes things important in leading life (health, education, cultural life etc.) while *capabilities* concerns individual combination of different functionings specific for human-beings or some group of people. So, social exclusion relies on inability of achieving certain 'functionings' or difficulties with reaching the goals which leads to deprivation and poverty - unwanted states in any society.

Multi-dimensional character of exclusion has been stressed by the following authors: De Haan [4], Bhalla and Lapeyre [6], Burchardt et al. [5] and Fisher [7]. Potential dimensions cover the different aspects: un/employment, markets (so difficulties with access to goods and services), neglecting of political laws and social relationships. Therefore in research conducted in this domain all aspects of exclusion processes should be examined not only individually but also from more general point of view. There are many intersections between the mentioned dimensions, for example: unemployment has the strong impact on poverty, poverty in turn causes limited access to services and products available on a market in local and global range and so on.

From individual as well as from society points of view the mentioned manifestations of exclusion processes lead to segregation of many sorts, sense of social inequality and conflicts in broader perspectives. Monitoring and investigation of the discussed phenomenon need to be performed continuously and be supported by information and communication technologies or better by specialized knowledge portals.

The described phenomenon basically relates to social exclusion which should be separated from voluntary exclusion – see Barry [8]. Exclusion of this sort is a specific one and not always is regarded as an unwanted process. On the contrary, as intentionally prepared and performed activities cannot be integrated with social exclusion.

The process considered as the solution to neutralize exclusion effects is called social inclusion. The social inclusion can be defined as ..."a process of improving the terms on which people take part in society" - see World Bank definition [15]. In sociology social inclusion means the provision of certain rights to all individuals and groups in society, such as employment, adequate housing, health care, education and training, etc. – see Collins Dictionary [12]. Of course social inclusion – as the process of organizing social life – sometimes is problematic or even inequitable – see Hickey and du Toit [9]. The concept of 'adverse corporation' brings better results because of its implementation in particular contexts – see [9].

Social inclusion as the process of counteraction of negative results of exclusion should be supported in many ways by information technologies. Social equality should be enforced by access to information and knowledge available in computer networks; M. Warschauer discussed many aspects of usability and consequences of technology and social inclusion intersection see [10]. In the next part importance of knowledge portals as natural source of information for modern society is presented.

3   Knowledge Portal for the Modern Society

Let us start discussion about portals from original definition of this term. The name – *portal* - comes from the portal architecture and denotes a door frame in the palaces, churches, town halls, etc. Portals were so meant to be gateways to the Internet, to the wide world for each user.

Portal - a website presenting the overview and systematic form the most important and best - developed articles, as well as other content related to the topic. It presents readers with the resources available on the subject in a more accessible and attractive than categories, indirectly encouraging the active involvement in the development of the content.

Portal - the website term, usually is seen as larger and more extensive than normal information or hobby pages. Internet portal, in addition to the information function, has various facilities for the user, such as: to customize the appearance of the page, the search engine websites also offer additional services, such as e-mailing or downloading files. Most portals encourages users to exchange opinion and discussion, allowing them to comment on articles published on them.

Web Portal (or Internet portal) - an online news service expanded to include a variety of features available, accessible from a single web address. The intention of the portal creators is to encourage users to set the address as the portal home page in a Web browser, and treat it as a gateway to the Internet.

Typically, the portal contains information of interest to a wide audience. As an example, you can specify the content of the website: sections of portals refer to current news, weather, websites, chat, discussion forum and information retrieval mechanisms.

Therefore we may call Web Portal - as an online information services extended to a variety of Internet functions, available from a single Internet address (compare [17]). Typically, the portal contains information of interest to a wide audience. In addition, web portals may offer free services such as email, web hosting.

Going further - Corporate Portal – can be defined as an integrated user interface, developing websites, and for the exchange of information, knowledge management in the company and the implementation of various business transactions. This is a point of access to all information resources and applications used. It integrates systems and technology, data, information and knowledge to function in the organization and its environment, in order to allow users a personalized and convenient access to data, information, knowledge, according to the tasks arising from their needs, anywhere, anytime, in safe manner and through a single interface. More advanced portals act as powerful workplace to ensure collaboration and exchange of documents. Figure 1 [24].

Fig. 1. Elements of Corporate Portals

C. Shilakes and J. Tylmann from Merill Lynch are recognized as creators of the Corporate Portals (CP) concept [21]. These portals should represent the following features [22]:

- integration of heterogeneous data, structured and unstructured, from the organization and its environment,
- integration of heterogeneous applications,
- providing information onto users, both automatically to authorized users and each of their request,
- portal interface can be adapted to individual needs,
- providing in-depth information and knowledge on very specific areas of individual users and groups of users,
- creating opportunities for communication, information sharing and collaboration between users or groups of users,
- categorizing data, information and knowledge available through the portal,
- publication and distribution of information and knowledge and its dissemination among employees.

The overall architecture of CP embraces four layers:

- presentation layer and personalization: defines how each user access to resources and services of the portal,
- taxonomy and search layer conditioning the ease and speed of getting the information sought,
- security layer: a very important element of the architecture. Portal must have built-in controls that effectively secure the data, knowledge, information, and applications from unauthorized access,
- integration layer: provides access to the data existing in the organization and its environment. Portal integrates applications and resources to facilitate access to them.

As a result we may formulate the following benefits of corporate portals:

- quick access to key information necessary for decision-making,
- reduce the time searching for the necessary information,
- access to knowledge regardless of the location of users, •
- good organization of work thanks to the integrated access to information.
- reduce the cost of distribution of information, mainly a reduction in spending on calls and paper.

More advanced version of CP is Knowledge Portal. "Knowledge Portal as a type of portal that purposely supports and stimulates knowledge transfer, knowledge storage and retrieval, knowledge creation, knowledge integration, and knowledge application (i.e., the processes of knowledge management) by providing access to relevant knowledge artifacts. Repository-oriented components and functionalities of a knowledge portal include a knowledge organization system, repository access, search, and applications and services. In addition to the repository-oriented functionality of a knowledge portal, such a portal must also offer network-oriented components and functionalities. Some types of knowledge are most readily transferred through direct interaction between a knowledge seeker and another knowledgeable individual. To that end, a knowledge portal also provides functionalities to identify and connect users based on their expertise, such as collaboration and communication tools" [25].

Knowledge Portal (KP) – can be defined as an online service that includes a generally reliable information about a specific fragment of reality and can be used in the further development of the issues or bind it with another issue. Knowledge expressed in many forms including its own theories is a crucial component of the discussed portals. The classic approach takes into account the following elements of KP:

- beliefs - judgments in the logical sense,
- justification - the belief is justified,
- veracity - the belief is true.

Typical examples of knowledge portal categories are:

- social sciences and humanities, such as astronomy, biology, chemistry, genetics, medical science, zoology,
- society, such as: anarchism, atheism, bible, biographies, philately, Hinduism, Judaism, Catholicism, the saints, the religious,
- geography, for example: individual continents, different countries, selected cities,
- national, for example, Poland, Germany, USA,
- culture - fiction, film, comics, art, games, anime,
- sports, for example, the Olympic Games, check, ski jumping, rallies,
- technique, such as: architecture, electronics, energy, computers, mobile phones, websites, army,
- social sciences and humanities, such as philosophy, history, psychology, sociology, foreign languages,

One of the oldest knowledge portals is presented in Fig. 2 (see: [19]). Main functionality of the portal concerns to co-operation with experts, e-learning offerings, Case Law Directory apart of typical FAQ (frequently asked questions) capabilities.

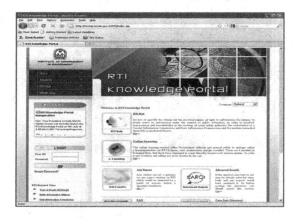

Fig. 2. An example of RTI Knowledge Portal

Next presented KP is ICSI Knowledge Portal (ICSI-KP) "a capacity building initiative of the Institute of Company Secretaries of India (ICSI) for its members and students is a reservoir of knowledge enabling the users access to huge pool of information including Bare Acts, case laws, notifications and circulars issued by Government and regulatory authorities from time to time" – description available from eJurix on the ICSI KP home page Fig. 3 [23].

Fig. 3. ICSI - Knowledge Portal

Another example of knowledge portal is depicted in Fig. 4 (available at [20]). The goal of its solution is to convince to company products as well as to offer complete courses devoted to accelerate user skills.

Fig. 4. Working with SONY Knowledge Portal

A useful tool for strengthening a network of people who are excluded can be a portal of knowledge. It serves not only as a tool to convey information, but above all to enrich and exchange their knowledge resources. Knowledge portal provides access to knowledge resources that are owned by all of the entities forming the network issues of excluded people. The first impulse to gain specific knowledge resources is to determine user's needs. Information goes to the portal, which locates the appropriate resource, a category of knowledge transfer process. Then the knowledge portal allows supplement domain knowledge. In this way, the knowledge portal becomes a compendium of knowledge on the particular matter. Site work requires not only knowledge of information technology, but appropriate organizational structure and knowledge management strategy. The proper functioning of the portal of knowledge brokers play an important role. They should be in the portals of knowledge as individuals or institutions performing oversight knowledge development in the portal. The main task of the broker is:

- assessment of the needs of the knowledge,
- location of resources,
- supervision of the transfer,
- assessment of the degree of absorption and utilization of knowledge.

All operators should be interested in the proper functioning of the portal of knowledge. Knowledge that remains in the resources of the organization is the basis for the acquisition of new resources. We have in this case to deal with a spiral of knowledge that all time develops, through which actors are able to develop. Basic knowledge resources of the company are based on five categories of knowledge (compare [18]):

- codified knowledge,
- explicit knowledge,
- knowledge protected,
- tacit knowledge,
- latent knowledge.

The portal should provide exploration of tacit knowledge and latent one. These two categories can contribute significantly to the development of the organization, and thus to achieve a competitive advantage. Portals should not only be source for the implementation of business functions typically associated with competitiveness, achieving measurable financial results. They should also be used for indirect actions that will achieve the objectives of a typical business. Different types of social organizations also need such solutions. A multitude of knowledge, which allows to solve problems of social institutions requires it to organize and codify the first stage of creating the portal. The next step is to complete this knowledge different solutions at the local level. The establishment and operation of such a portal will allow institutions to function more efficiently to deal with the problems of excluded people. Perhaps those who are excluded will return to society. However, their efficiency and effectiveness will be determined only by time.

4   A General Concept of Knowledge Portal for Exclusives

Highly specialised knowledge portal will be presented through description of two portal elements: assumptions and functionality. In the last part KP architecture will be demonstrated in graphical form.

Functionality. Designing the fundamental functionality of the portal for exclusives we have to take into consideration dimensions of the exclusion mentioned above. The used cases should be aggregated into following groups:

- support for enter/return to the labour market,
- goods and services flow targeted to exclusives,
- communication and content management features.

The features focused on the help for the unemployed have to fulfil the following goals:

- Gather and present job offers that may be beneficial but also accessible by exclusives. There will be possibility for entering such offers directly into portal but also the portal will be loaded automatically with suitable offers from other portals specialized in job offering.
- Gather and present profiles of people looking for a job.
- Automatically match the employers and employees and suggest corresponding job offers for candidates and candidate's profiles for employers.
- Encourage exclusives to register their profiles and to make an effort to find or change a job.
- Enhance skills and competencies of candidates by e-learning features.
- Encourage employers to give a chance to exclusives by showing them benefits of hiring exclusives.

The next group of portal's features will have the following goals:

- Collect offers of goods or services directed to exclusives that are available by no charge or by a very low prize. Such offers will be entered directly to portal but also will be searched in other trade portals.
- Present the announcements of exclusives specifying the goods or services that they require.
- Match the necessitous with the offers.

In both those groups of functionalities we have to remember that the exclusives may use the portal directly or be represented by the people engaged in the process of help for exclusives such as social workers. We also take into account that the personal data of exclusives must be protected with special care.

The last mentioned part of functionalities deals with the educational and communicational features directed to:

- Politicians, journalists, researches and others that deal with the problem of exclusives and want to learn about problems or exchange ideas.
- Exclusives – to be aware of the law or the programs dedicated to help them.

Assumptions. We assume the following groups of users are the target users of proposed portal:

- Policy makers – represented by politicians and officials who are responsible for legal solutions, law creation and adjustments.

- Social workers employed in:
  - Governmental/local governmental units.
  - NGO's – Non-governmental organizations.
- Institutions interested or engaged in the exclusion problem like universities/scientific organisations and their researchers, media and journalists.
- Independent entities which are interested in the exclusion problem.
- Commercial enterprises which develop programs/solutions targeted to help excluded groups.

By using another perspective we can divide the users into:

- Corporate users representing authority organisations.
- Individuals representing basically themselves that consist of:
  - Authorities that should be verified.
  - Regular users.

The main basis of the portal assume that every published content is available for all portal users (except from the personal data of exclusives). There will be no confidential matters stored in the portal so there is no need to build sophisticated information protection module.

No confidentiality assumption determines the simplicity of permissions model in the portal. The administrators of the portal will be able to configure it's modules and manage the permissions for every defined role, but there is no need to limit the access to the specific content.

We perceive that if we build a dedicated authentication module with separated credential management it will become a serious pain point for the users who will be forced to create and remember yet another user login and password. That is why we would like to integrate with as many as possible authentication providers that can exchange the data with the portal. Such approach will allow the users to use in our portal the same credentials as they are using in their enterprise systems or social solutions. Our portal will be integrated with:

- LDAP solutions,
- Facebook/twitter/google accounts.

Our authorisation module architecture approach will allow organisational users to authenticate in our portal with the corporate login and password or even include the portal into the corporate SSO (single sign-on) solution. Individual users will be able to access the portal with the private facebook account and sign in automatically.

Of course we can use the mixed mode of authentication which means that corporate users can use also their private social accounts if they prefer or if the integration with the corporate LDAP will be impossible.

The corporate users of the portal should be verified by:

- Portal administrators,
- Organisation representatives.

In case of the individual users that should be verified – this task will be done by portal administrators.

One of the most important assumptions regards the functional offerings of the portal. It is not designed to work with specific excluded people and solve their particular problems. The main aim of the portal is to connect, enhance the interchange of ideas, knowledge and experience between the experts engaged in exclusion problem. It is designated to inspire all the stakeholders to solve the general problem of exclusion and provide them with relevant information and knowledge. That is why the portal will not consist workflow functionality/application processing capabilities but it is focused on information and knowledge tools processing, ideas management, communication platform and e-learning modules.

Architecture. We propose the Knowledge Portal for Exclusives (KPE) to be built in traditional 4 layers architecture:

- Integration layer – build to provide communication with the other systems collaborating with our portal. They can be:
  - Authentication provider like google, facebook, twitter, MS and others
  - Content delivery portals belonging to governmental organisations or similar initiatives focused on exclusion problem
  - Commercial/business portals publishing jobs offers and portals providing consumer to consumer & business to consumer sales services.
- Database layer – build as relational database storing the business and process objects. In this layer we will store also all documents, multimedia files, learning objects that has no relational structure. This layer will also provide the services of reporting and integration with other systems.

Database layer will be supported also with Knowledge Base. The knowledge to the Knowledge Base will be supplied by:

- Experts (portal users)
- Knowledge exploration module that will be operated by portal power users with data exploration skills.

- Application layer that will be responsible for whole business logic. It will provide:

- the basic portal functionality,
- the communication services that will give the users possibility for on-line bilateral communication, teleconference services, off-line communication,
- e-learning services with all capabilities for hosting and providing on-line courses,

- Presentation layer build in portal technologies (which means that all the portal features will be accessible by web browser). We must take into account that for some users the portal will not be a tool in which they will work every day. That is why from user perspective it is crucial to provide them notifications of all important event happening in portal as new available content, tasks for users, activities expected to be done by users and others. Such notifications will be provided by (Fig. 5):

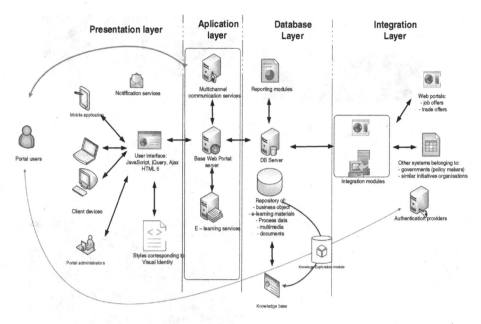

Fig. 5. Architecture of knowledge portal for exclusives

- Automatically generated emails,
- Mobile application push notifications,
- Newsfeed published by social portal as Twitter or Facebook.

Presentation layer will provide end users access to the information and knowledge stored in the database layer with regard to: users permissions and users preferences (content should be targeted to end user's needs and requirements).

5   Implementation Examples of Knowledge Portals for Exclusive in Local Governments

Presented earlier general knowledge portal for exclusives can be implemented at all levels of the modern society. We focus on regional implementation of KPE in Silesia agglomeration. There are some "sectors" on existing portals devoted to exclusion problems (see Fig. 6).

The problem of supporting exclusives is very important in Silesia region. During transformation starting at the beginning of 90's in Poland many miners and workers of heavy industry lost their jobs. So the problem became very important also for local governments. The content of the selected Silesian cities portals: Jastrzebie Zdroj, Wodzislaw Slaski, Gryfow Slaski oraz Miasteczko Slaskie reflects the typical approach to "excluded" people; actually on these portals habitants may find basic information

Fig. 6. Portals of Silesia agglomeration with exclusive quests

about exclusive phenomena presenting some initiatives and projects. Therefore the idea of development KPE seems to be very natural.

The presentation of general assumptions and specialty such of portals we start from itemizing of potential users. At least the following group should be included:

- Knowledge bidders - e.g. institutions responsible for the creation of legal solutions, laws and regulations, NGOs, institutions interested or involved in the problem of exclusion (universities, research institutions, scientists, the media, journalists), independent entities interested in the problem of exclusion.
- Individual users, - e.g. citizens of that region who are currently subject to marginalization or exclusion, and those seeking to achieve social equality.
- Social workers, commercial enterprises that develop software solutions targeted to excluded groups and individual experts.
- Experts, associated with the preparation of educational materials and the development of e-learning courses for end users platform.

Defining technological assumptions and considering general contexts we expect from the discussed platform:

- Full integration, assuming that in the basic system infrastructure - service target group (in this case those who are vulnerable or excluded) relies on integration of heterogeneous information generated by specific entities within e.g. A single portal and using the presentation layer respectively made available to specific groups of customers.
- Groupware approach which is based on intelligent infrastructure, critical systems and cooperating users, helping to improve efficiency in the pursuit of counteracting exclusion or marginalization.
- Applying of the comprehensive contact covering implementation of smart technology management and public information resources, allowing to systematize and increase citizens' access to information and knowledge, concerning measures to counteract exclusion and marginalization phenomena using ICT infrastructure.

As a result we may expect the following benefits of KPE at the local level:

- increasing the number and quality of habitants's contacts with the public,
- building a communication network between the public sector and people,
- reducing the sense of isolation and loneliness among the excluded,
- integration of existing information systems,
- adaptation of persons excluded or marginalized to benefit from services that enable increase their mobility and social integration of access to ICT solutions, health care in order to develop greater independence and autonomy from third parties,
- reducing social exclusion phenomena.

6   Conclusion and Further Research

One of the most important problems of the modern society at the local, national and global levels is exclusion processes. The next findings can be formulated from the research:

- phenomena exclusion and inclusion in society are complex and multidimensional. Therefore investigation of such processes is difficult also because of differentiation of components, approaches and dimensions,
- information and communication technologies must support all processes belonging to registration and all services typical for exclusion and inclusion phenomena in modern society,
- the best solution for the discussed problems is specialized knowledge portal – proposed in the paper as a form of an initial version. At the beginning architecture will be developed for some narrow application and systematically extended covering new areas and levels of supporting.

Further investigations can be devoted to more deeply analysis of the problem in order to fulfill requirements of knowledge portal users. For example specification of knowledge bases and courses essential at local and/national level of administration.

References

1. Silver, H.: Social exclusion and social solidarity: three paradigms. Int. Labour Rev. **133**(5–6), 531–578 (1994)
2. Saith, R.: Social exclusion: the concept and application to developing countries. In: Stewart, F., Saith, R. Harriss-White, B. (eds.) Defining Poverty in the Developing World, Palgrave, pp. 75–90 (2007)
3. Sen, A.: Social Exclusion: Concept, Application, And Scrutiny. 'Asian Development Bank, Manila (2000)
4. De Haan, A.: Social Exclusion: Towards an Holistic Understanding of Deprivation. Department for International Development, London (1999)
5. Burchardt, T., Le Grand, J., Piachaud, D.: Introduction. In: Hills, J., Le Grand, J., Piachaud, D. (eds.) Understanding Social Exclusion. Oxford University Press, Oxford (2002)
6. Bhalla, A., Lapeyre, F.: Social exclusion: towards an analytical and operational framework. Dev. Change **28**, 413–433 (1997)
7. Fischer, A.: Reconceiving Social Exclusion, BWPI Working Paper 146, Brooks World Poverty Institute, Manchester (2011)
8. Barry, B.: Social Exclusion, Social Isolation and Distribution of Income. Centre for Analysis of Social Exclusion, London School of Economics, London (1998)
9. Hickey, S., du Toit, A.: Adverse Incorporation, Social Exclusion and Chronic Poverty, Working Paper 81, Chronic Poverty Research Centre, University of Manchester (2007)
10. Warschauer, M.: Technology and Social Inclusion: Rethinking the Digital Divide. The MIT Press, Massachusetts Institute of Technology, Cambridge (2004)
11. Rawal, N.: Social inclusion and exclusion: a review. Dhaulagiri J. Sociol. Anthropol. **2**, 161–180 (2008)
12. Social Exclusion portal: http://www.gsdrc.org/go/topic-guides/social-exclusion/definitions-and-different-understandings-of-social-exclusion. 23 April 2014
13. Exclusion-Inclusion in Suburb: - http://www.hioa.no/eng/About-HiOA/Centre-for-Welfare-and-Labour-Research/NOVA/NOVA-Projects/Prosjekter-migrasjon-og-transnasjonalitet/Avsluttede-prosjekter/2011/Exclusion-and-inclusion-in-the-suburb/(language)/eng-GB. 23 April 2015
14. H2020 sections: http://ec.europa.eu/programmes/horizon2020/en/h2020-sections. 23 April 2015
15. World Bank website: http://www.worldbank.org/en/topic/socialdevelopment/brief/social-inclusion. 23 April 2015
16. Collins Dictionary website: http://www.collinsdictionary.com/dictionary/english/social-inclusion. 23 April 2015
17. Internet portal - http://en.wikipedia.org/wiki/Internet_portal. 23 April 2015
18. Knowledge concept - http://en.wikipedia.org/wiki/Knowledge. 23 April 2015
19. RTI Knowledge Portal - http://rti.img.kerala.gov.in/RTI/index.jsp. 23 April 2015
20. Sony Knowledge Portal - https://training.sony-europe.com/. 23 April 2015

Knowledge Management in Distributed Agile Software Development Projects

Mohammad Abdur Razzak[1]([✉]), Touhid Bhuiyan[1], and Rajib Ahmed[2]

[1] Department of Software Engineering, Daffodil International University,
Dhaka, Bangladesh
razzak@live.se, t.bhuiyan@diu.edu.bd
[2] Exertis Ztorm, Stockholm, Sweden
l.rajibahmed@gmail.com

Abstract. Knowledge management (KM) is essential for success in global software development. Software organizations are now managing knowledge in innovative ways to increase productivity. In agile software development, collaboration and coordination depend on the communication, which is the key to success. To maintain effective collaboration and coordination in distributed agile projects, practitioners need to adopt different types of knowledge sharing techniques and strategies. There are also few studies that focus on knowledge sharing in distributed agile projects. This research investigates the knowledge sharing techniques and strategies applied by the practitioners in distributed agile projects. In addition to that, challenges faced by the practitioners during knowledge sharing in distributed agile projects are also identified and discussed.

Keywords: Knowledge management · Knowledge sharing · Distributed · Agile · Global software development

1 Introduction

Software engineering is a knowledge intensive area. This forces software organizations to manage their knowledge and later use it in smarter, innovative ways to solve problems (Schneider 2009). It helps software development organizations to acquire and maintain a competitive advantage. KM is crucial for success in global software development (Richardson et al. 2009).

Global software development can be described as "software work which is attempted in different geographical locations across the national boundaries in a coordinated fashion, to involve synchronous and asynchronous interaction" (Sahay et al. 2003). Software developers work with knowledge and are dependent on each other's work. In global software development this synchronization is dependent on KM. Some studies have identified that knowledge sharing is difficult in distributed agile project due to the lack of face-to-face communication between team members (Boden and Avram 2009; Holz and Maurer 2003). In the agile software development collaboration and coordination depends on communication, which is crucial to successful software development (Šmite et al. 2010).

© IFIP International Federation for Information Processing 2015
E. Mercier-Laurent et al. (Eds.): AI4KM 2014, IFIP AICT 469, pp. 107–131, 2015.
DOI: 10.1007/978-3-319-28868-0_7

One of the major objectives of KM is to improve productivity through effective knowledge sharing and transfer (Kavitha and Ahmed 2011). So, the success of agile projects relies on effective knowledge sharing among teams.

This research focuses on exploring knowledge sharing in distributed agile projects. More specifically, this research attempts to identify knowledge sharing techniques, strategies and practices that take place between locally and globally distributed agile teams, and the challenges faced by the practitioners in a distributed agile environment. We are driven by the following research questions:

RQ1: How do team members contribute to knowledge creation in a distributed agile project?

RQ2: How do team members share knowledge in a distributed agile project?

RQ3: What are the challenges faced by the practitioners when sharing knowledge in a distributed agile project?

2   Related Work

Software development is considered to be a complex, knowledge intensive and rapidly changing activity, where a number of individuals, teams and organizations are involved in fulfilling common goals, interests and responsibilities (Curtis et al. 1988; Nicholson and Sahay 2004). Technological and strategic knowledge helps developers to communicate; so it is essential to keep the knowledge stored in the organization for the future reuse. Davenport and Prusak (Davenport and Prusak 2000) define it as "a method that simplifies the process of sharing, distributing, creating, capturing and understanding the company's knowledge". As the size of the organization grows rapidly, it becomes harder to find out where the knowledge resides. Research shows that if the companies manage their knowledge in a better way, they can increase quality, and decrease the time and development costs (Rus et al. 2002). To improve the organizational performance, it is important to manage knowledge in a structured way which will help to convey the right knowledge to the right people at the right time. O'Dell and Grayson (O'Dell and Grayson 1998) discussed that, knowledge management is not a vital methodology; it is a framework, a management mind-set which is based on past experiences and the creation of new wheels for exchanging knowledge.

To foster dynamic knowledge sharing, improve productivity and coordination in software development teams, agile approaches were introduced. Agile teams share knowledge through several practices (Chau and Maurer 2004): pair programming, release and sprint planning, customer collaboration, cross-functional teams, daily scrum meetings and project retrospectives. But, the authors (Chau and Maurer 2004) argue that, these practices are team-oriented and rely on face-to-face interaction between team members. These practices do not facilitate knowledge sharing in distributed agile teams but are effective for collocated and small teams. In one study Dorairaj *et al.* (Dorairaj et al. 2012) reported that in distributed agile project, team members practice sprint planning, daily scrums, sprint reviews and project retrospective meetings. Distributed agile team members share knowledge through effective use of knowledge management tools like *Wiki*, pair-programming and video-conferencing.

Michael Earl (Earl 2001) has classified knowledge management into three categories: technocratic, economic and behavioral. Earl also divided these three categories into seven schools, Technocratic: *Systems, Cartographic and Engineering*, Economic: *Commercial* and Behavioral: *Organizational, Spatial and Strategic*. Both *codification* (Hansen et al. 2005) strategy and *systems school* practice depend on the technology which applies Nonaka's (Nonaka 1994) *externalization* conversion technique to convert tacit knowledge into explicit knowledge. Research shows that the technocratic school is closely related with traditional software development and those who are developing software through traditional approaches they are probably benefiting from the technocratic schools (Dingsøyr et al. 2009). On the other hand, behavioral schools are more related with the agile approaches and agile teams are more benefit more from the behavioral school. A survey in traditional and agile companies shows that agile companies seem to be more satisfied with their knowledge management approaches compared to traditional companies (Bjørnson and Dingsøyr 2009). In agile software development, knowledge sharing happens through the interaction. Developers share knowledge by working together and through close interaction with customers; and more specifically, pair programming, extreme programming, daily scrum meetings, and sprint retrospectives in Scrum. In traditional software development, knowledge management relied primarily on explicit knowledge but in the agile software development KM relies on tacit knowledge (Nerur et al. 2005). In agile software development, information radiators and collocating teams are related with the spatial school (Bjørnson and Dingsøyr 2009).

In traditional software development, knowledge stored explicitly in the documentation, but in the agile development methodology the knowledge is tacit (Kavitha and Ahmed 2011). Extracting tacit knowledge to create explicit knowledge is one of the greatest challenges of knowledge organization (Nonaka and Konno 1998). Due to the absence of explicit knowledge in the agile software development, experts need to spend much of their time on repeatedly answering the same questions, knowledge is lost when experienced developers leave project, there is less support for re-usability and there is less contribution to organizational knowledge (Kavitha and Ahmed 2011). In the agile collocated development, informal communication is the key enabler for knowledge sharing but when an agile project is distributed, informal communication and knowledge sharing is a challenge due to low communication bandwidth as well as social and cultural distance (Maalej and Happel 2008). Due to spatial, temporal and cultural factors, communication also becomes aggravated in the distributed settings (Hildenbrand et al. 2008). Several studies (Holz and Maurer 2003; Boden et al. 2009) also point out that, knowledge sharing in the distributed agile projects is difficult due the challenges in communication, especially face-to-face interaction between team members in different geographical locations. To address these problems, we investigated how shared knowledge creation and transfer activities performed in the distributed agile projects. Along with that, we also investigated what challenges are faced by the practitioners when sharing knowledge among globally distributed agile team members.

3 Method and Data Analysis

Because this research addresses an issue *"How can we retain the benefits that agile practices provide with respect to KM in distributed agile projects"* which is rather under-investigated, this study takes an explorative approach. Exploratory research helps to find out what is happening, seeking new insights and gathering ideas (Marczyk et al. 2010; Runeson and Höst 2009). In some qualitative research, data collected through observation or interviews are exploratory in nature. So, extensive interviews are helpful to handle this type of situation (Sekaran 2006). This type of exploratory research was also helpful in achieving our goal through analysis of similarities and differences among the cases (Creswell 2008). The primary focus of this study was to discover the knowledge sharing activities in distributed agile projects in order to identify techniques, strategies and challenges.

3.1 Sampling

The selection criteria for these interviewees were based on the kind of company they work at, the experience of the company in distributed agile development (more than 2 years), interviewee role in the distributed team as well as in the company, project duration and project distribution. The participants of this research were project managers, team leaders, software architects, line managers, senior software developers, system developers and Scrum masters in different countries involved in distributed agile projects, located in different countries i.e. Sweden, Norway, Germany, Ukraine, China, India, Bangladesh, USA, and Latvia. To get the rounded perspective of this research phenomenon we included different roles from the agile team.

3.2 Data Collection

There are three types of interview techniques namely structured, semi-structured and unstructured (Flick 2009). Due to the qualitative nature of this study we used semi-structured interviews for conducting a series of interviews in software industries involved in distributed agile projects. According to Robson (Robson 2002), an in-depth semi-structured interview is helpful in *finding out what is happening and seeking new insights. Seventeen* semi-structured interviews were conducted from *seven* teams in order to identify how practitioners are creating, storing and sharing knowledge related to software development among geographically distributed agile teams. These semi-structured interviews were a combination of both open and focused questions. It helps both interviewer and interviewee to discuss a topic in more details. Before the interviews started, we discussed about overall goal of this research to interviewee. The interview questions were *descriptive* and with the base questions there were follow up questions asked based on the discussion. We were concern about some key terms: *shared knowledge creation, knowledge transfer, strategies and challenges* which later helped us for data analysis and those terms which also evolve with interview questions.

We conducted seventeen semi-structured interviews from six different companies. The selected companies are involved with software product development, have different organizational settings and structure and are located in different countries. The duration of these interviews averaged 60 min and the interview sessions were tape recorded. Among the seventeen semi-structured interviews, nine were conducted through Skype and eight were face-to-face, depending on distance between interviewer and interviewee.

3.3   Analysis and Synthesis

In qualitative research, data analysis is the most difficult and crucial aspect due to raw data sets. According to Basit (Basit 2003), raw data can not help the reader to understand the social world or the participants view unless such data is systematically analyzed. To organize collected data we adopted *thematic analysis* (Braun and Clarke 2006) technique during analysis. Thematic analysis is used to identify, analyze and report patterns or themes within data. It minimally organizes and describes data set in detail. In thematic analysis a theme captures data with relation to research questions and represents them in a pattern within the data set (Braun and Clarke 2006). This analysis is performed through a process which maintain six phases to establish meaningful patterns of the data set. Braun and Clarke (Braun and Clarke 2006) provides an outline through the six phases of analysis. These phases are: familiarization with data, generating

Table 1. Overview of the studied distributed Agile projects

Projects	Project distribution	Team size	Team types	Agile position/roles
Alpha	Sweden-Germany	6–7[a]	Dispersed	Team Leader
				Developer
Beta	Norway-Bangladesh	5–6[a]	Dispersed	Project Manager
				Developer
Gamma	USA-Bangladesh	12–16[b]	Distributed	Head of Engineering
				Senior Developer
				Developer
Delta	Sweden-Bangladesh	16–18[b]	Dispersed	Software Architect
				Developer
Epsilon	Latvia-Ukraine	11–15[b]	Distributed	Project Manager
				Developer
Zeta	Sweden-China	26–35[c]	Distributed	Line Manager
				Software Developers
				System Developer
Eta	Sweden-India	45–55[c]	Hybrid	System Developer
				Scrum Master

In Table 1, [a], [b], [c] indicates Small, Medium and Large scale teams respectively

initial codes, searching for themes among code, reviewing themes, defining and naming themes, and producing the final report.

In the *first* stage, we transcribed all the collected interview data into written form in order to conduct a thematic analysis. It helped us to identify possible themes, patterns and to develop potential codes (Guest et al. 2011). *Second* phase started with initial codes from the extracted data. There are different types of Coding techniques suggested in different studies such as; *open, axial, selective, descriptive/topic and pattern or analytic* (Miles and Huberman 1994; Punch 2009; Shull et al. 2007). In our case, we applied open coding technique and went through all transcribed textual data by highlighting sections of the selected codes. That also helped us to relate coded data with research theme and research questions. In *third* stage, we analyzed broader level of theme rather than codes that helps to sort different codes into potential themes (Braun and Clarke 2006). As Braun and Clarke suggested coding as many potential themes/patterns as possible because initially some themes seems to be insignificant, but later they may be important in the analysis process. Later, mind mapping tools were used to represent them into theme-piles. This stage gave us a sense of the significance of individual themes. Stage *four* is reviewing themes. In this stage we identified irrelevant (not enough or diverse) data with relate to different themes and broken down into separate themes. After refining all themes we identified "essence" of each theme and different aspects of the data each theme captures in stage *five*. At the end, in stage *six*, we provided extract data with relate to research questions and present some dialog that connected with different themes in support of results and discussion sections.

4   Validity Threats

To handle validity threats it is important to identify all possible factors that might affect the accuracy or dependability of the results.

4.1   Internal Validity

Internal validity for qualitative research mostly relates to the researchers biasness and interpretation of data (Bleijenbergh et al. 2011). For finding a similar knowledge level for our interviewees, we went on interviewee profiles on *Linkedin* and their years of experience. After finding out the basic information, the interviewer sent a formal email to the interviewee with an invitation letter about becoming involved with this research. To mitigate the threat of following our own bias, interview questions were designed to have a majority of open ended questions. Every interview started with a similar introduction and some clarification questions. Then the recorded interview was transcribed immediately afterward to reduce the risk of missing some information. Furthermore, researchers sent an interview report to the interviewee in order to check whether interview data was correctly transcribed and to confirm the content indicated participants thoughts, viewpoints, feelings and experiences. In qualitative research it is important to understand the interviewee's inner meaning words. To maintain reliability during data analysis we used a thematic, qualitative data analysis technique, that helped to identify, analyze and report themes within data. The extracted data from the transcribed data was checked twice for any discrepancy by two researchers.

4.2 External Validity

External validity threat is more applicable to research that are quantitative and which tries to generalize outcome of the research. However, our findings can be generalized only for the agile software development teams which are involved in the development of a shared project from distributed locations.

5 Results

In this section, we describe different findings (techniques, strategies and challenges) from the *seven* cases, that promote effective knowledge creation and sharing activities in distributed agile projects.

5.1 Knowledge Creation: Locally and Globally

We have found that distributed agile project teams practice different types of techniques for both local and global shared knowledge creation. *Pair programming, customer collaboration, Scrum/Kanban* boards and *community of practice* are explicit practices used by the teams to perform both local and global shared knowledge (see in Table 2).

Table 2. Knowledge creation techniques: Locally and Globally

Techniques	α	β	γ	δ	ϵ	ζ	η
Pair programming	L,G	L,G	L,G	L,G	—	L	L,G
Customer collaboration	L,G	—	L,G	L	L	L,G	L,G
Scrum/Kanban boards	L	—	L,G	L	—	L	L,G
Innovation boards	—	—	—	—	—	—	L,G
Workshops/Seminars	—	—	—	—	—	L	L
Community of practice	—	—	—	L,G	L,G	L,G	L,G
Technical presentation	—	—	L	—	—	L	L
Technical forum	—	—	—	—	—	L,G	L,G

In Table 2, L indicates Locally, G— Globally and "—" not in practice. Dispersed teams- α, β, δ; *Distributed teams-* γ, ϵ, ζ; *Hybrid team-* η

5.1.1 Pair Programming

Pair programming is used for both local and global knowledge creation. From the series of semi-structured interviews we have found that both local and remote team members work together in one workstation to solve specific problems. They help each other to share their thoughts and create knowledge through discussion. In two cases, we have found that teams do not perform pair programming for shared knowledge creation among remote team members. In the Epsilon (ϵ) project, all development team members are in one site, and for that reason they do not need to perform pair programming for global shared knowledge creation. However, Zeta (ζ) project is a collaboration with a Chinese team on the same product, but the development team does not have any dependency. The development teams working on different modules and later core developers merge all modules together for specific release. But the local teams in Zeta (ζ) project perform pair programming.

5.1.2 Pre-planning Game/Customer Collaboration

In the development cycle the customer has an important role. Customer collaboration helps teams to build up technical-business collaboration on a project and also helps to set the direction of the project. In agile software development customers are always involved with the development teams by providing project requirements and performing acceptance testing. Through customer collaboration agile teams participate in creating local knowledge. Evidence was also found from different cases that customers are also involved with the remote development teams to create knowledge through continuous discussion and features feedback. We have also found that customers are involved in issue tracking systems, which helps both the project manager and the developers towards early iteration. In two cases (δ and ϵ), we found that customer collaboration performed only in the local sites for shared knowledge creation.

5.1.3 Scrum/Kanban Boards

The are two types of boards used by the office to create knowledge and common understanding. A *Scrum board* is used for teams that plan their work in sprint. A *Kanban board* is used to manage and construct team work in progress. In Table 2 it is shown that, teams use *Scrum* and *Kanban* boards for shared knowledge creation among both local and globally distributed team members. In two cases (γ and η), we found that teams are using boards both locally and globally. In Gamma (γ) project, the remote team has a sub-Scrum board, which is replica of the main Scrum board. Along with that, the local team (in γ project) upload pictures of the main Scrum board into a repository every day. But in Eta (η) project, teams use a visual Scrum board to perform shared knowledge creation among distributed team members.

5.1.4 Innovation Boards

Most innovative ideas are kept in the human mind as tacit knowledge. Due to continuous work loads, sometimes it is impossible to have a discussion with a team member, or other knowledgeable person. So, rather than talking with someone, people share their ideas through the innovation board in an explicit way. In one interview the researchers found that teams are using innovation boards to share their ideas with both collocated and remote team members.

5.1.5 Workshop/Seminars

Weekly or monthly workshops and seminars are arranged through collaboration between business teams, technical teams and customers, in order to share knowledge about projects and the latest technologies. This kind of workshop facilitates common understanding and communication between different team members. Workshops also help to facilitate tacit knowledge sharing through *socialization*. In the studied cases, we only observed *large-scale* teams practicing these techniques locally, to create shared knowledge. Later, the theme of the workshops/seminars was shared among remote team members through repositories.

5.1.6 Community of Practice

To succeed in agile projects, learning is an important asset for agile teams. Agile teams practice two modes of learning: *peer learning* and *community learning*. In *peer learning*, team members start learning through interacting and collaborating with team members. *Community learning* is accessing and conceiving information that is available in knowledge archives or in discussion forums. We found community of practice within different projects, where it performed to share knowledge creation among local and remote team members.

We also found that to create shared knowledge, teams perform *technical presentations*. But these activities are only performed in the local site and later slides or documents are shared among remote team members. Technical forums are also in practice to perform shared knowledge creation between local and remote team members.

5.2 Knowledge Sharing: Locally and Globally

Knowledge exchange is always challenging in distributed agile teams due to a lack of face-to-face interaction among team members. Practitioners and researchers are trying to mitigate these challenges by initiating different kinds of techniques and tools. From the studied cases we observed that practitioners maintain different types of tools and techniques to share knowledge among globally distributed teams. Based on the findings, these knowledge sharing techniques are listed in Table 3.

All studied projects are concerned with using repositories to share knowledge between local and remote team members. Most of the task and product related knowledge is kept in the repositories, which are easy to access by the remote team

Table 3. Knowledge sharing techniques among different sites

Techniques	α	β	γ	δ	ϵ	ζ	η
Repositories	L,G	L,G	L,G	L,G	L,G	L,G	L,G
Pair programming	L,G	L,G	L,G	L,G	—	L	L,G
Version control	—	—	—	—	L,G	—	—
Screen sharing	G	G	G	G	—	—	—
Daily scrum	L,G	—	L,G	—	L,G	L	L,G
Weekly sprint status	L,G	—	L,G	L,G	G	L	L,G
Common chat room	—	—	L,G	L,G	L,G	L,G	L,G
Technical forum	—	—	—	—	—	L,G	L,G
Discussion forum	—	—	L,G	L,G	—	L,G	L,G
Electronic board	—	—	—	—	—	—	L,G
Online conference	G	—	G	G	G	L,G	L,G
Rotation/Visit	—	—	—	—	G	G	G

In Table 3, L indicates Locally, G— Globally and "—" not in practice. Dispersed teams- α, β, δ; *Distributed teams-* γ, ε, ζ; *Hybrid team-* η

members. Different teams also depend on daily scrum, weekly sprints status, discussion forums, online conferences and common chat rooms to share knowledge between local and remote team members. Electronic boards are helpful for sharing knowledge across remote teams. Only one case was found where knowledge is shared between both collocated and distributed teams through electronic boards.

5.2.1 Repositories

To share knowledge among distributed sites, local teams used different types of repositories like *Wiki, JIRA, Redmine, Confluence and GitHub etc.* These types of repositories provide efficient mechanisms to access codified knowledge. From the gathered data it is evident that (see in Table 3) practitioners are most dependent on repositories to share knowledge among both local and distributed team members.

Wiki, according to Ulrike Cress (Cress and Kimmerle 2008), provides new opportunities to learn and use collaborative knowledge building and sharing, through social interaction and individual learning. In different cases we found that *wikis* are helpful for starting new threads and discussing issues with other team members. It is also helpful for new team members as it (the *wiki*) provides detailed information about features, documents and so forth.

Project and Issue tracking: Nowadays, almost all medium and large distributed or dispersed agile teams are using *JIRA/Redmine* to track issues, bugs, tasks, deadlines, codes and hours. As collaboration and content sharing tools practitioners used *Confluence* to share docs, files, ideas, specifications, diagrams and mockups. During the interview one project leader said,

...Most of the time we share tacit knowledge between both local and global teams. After that, the information is converted through Redmine to make it explicit... **Project Leader - Alpha project**.

It is also evident that, in one case we found local teams upload *Scrum* board pictures, slides and workshop information in the repositories, which is codified and easy to access by the remote team members.

5.2.2   Pair Programming

Pair programming plays an important role in creating and sharing developers knowledge in both locally and globally distributed project. In pair programming, two developers work together at one computer with a common goal (Palmieri 2002). In the studied projects, we found teams are using pair programming techniques to share knowledge among remote team members. Team members use Skype to share screens among remote team members. Along with that we also found that teams use *TeamViewer* and *VPN* services to share the same computer screen with remote team members, in order to perform pair programming.

5.2.3   Daily Scrum/Weekly Sprints Status/Online Conferences

Scrum meetings are a source for sharing project progress information among team members. Usually a Scrum standup meeting is held in collocation. From the gathered data we found that distributed teams practice Scrum standup meetings with Internet Relay Chat (IRC), Skype or other group chatting software. Through daily Scrum weekly sprints status/online conferences local teams share knowledge with distributed team members. In one case (ζ), we found that due to less dependency, the development team does not need to perform Scrum meetings/weekly sprint status. But team (ζ) maintain online conferences in order to share knowledge among remote team members (if needed). However, apart from Beta (β) project, other projects explicitly maintain *online conferences* globally for knowledge sharing among distributed team members.

5.2.4   Common Chat Room

Common chat rooms are useful for exchanging knowledge among distributed teams. From the empirical findings we observed that for faster and quicker communication among distributed team members, *medium-* and *large-scale* teams maintain common chat rooms.

In one case, a software architect said that *...the Sprint management system handles all task related knowledge but for the domain related knowledge sharing we maintain a common chat room, which helps us to resolve specific problems within a short time* - **Software Architect, Delta project**.

But, in another case we found that, it is not an efficient way to communicate among distributed teams due to language barriers, common understanding, technological factors and so forth. Frequently misunderstandings occur and things go wrong. To mitigate these types of problem, practitioners also suggested different types of mitigation techniques.

5.2.5   Technical Forum

The idea behind a technical forum is *learning through sharing knowledge*. Technical forums are like communities of practice which create a network between technical team members. They are self-organizing groups that consist of individuals who share information, experience and technical skill on a specialized discipline (McDermott 1999). Technical forums assist distributed teams in quick problem solving and reduce development time since team members do not get stuck on recurring issues. Building trust between team members in the distributed environment is challenging; so knowledge sharing through technical forums can build trust between developers. Technical forums help to create and share both local and distributed knowledge. We have found that large-scale team members practice *technical forum* techniques to share knowledge among remote team members.

5.2.6   Electronic Board

The office boards hold a lot of knowledge which is difficult to share among distributed teams. The interviews revealed that practitioners are using electronic boards to share and access knowledge both locally and globally. Electronic boards hold the tasks list to perform, latest information and along with that necessary technical and business information are regularly updated in a wiki. Electronic boards help to decrease the communication overhead.

In one case an interviewee said ...*I am not satisfied with the current tools; Its tough to describe designs to new team members. Visual aids are helpful during discussions* - **Project Manager, Beta project.**

5.2.7   Rotation/Visits

The primary intention of team member rotation between different sites is knowledge sharing. Due to frequent face-to-face interaction with product owners, on-site team members get more business and domain related information than offshore team members (Sureshchandra and Shrinivasavadhani 2008). A lack of face-to-face meetings and poor socialization also causes a lack of trust among distributed team members (Moe and Šmite 2008). Rotation between on-site and distributed team members promotes the sharing of business and domain related knowledge across the teams. From the data gathered we found that, both *distributed* and *hybrid* teams visit remote sites and rotate team members to increase the trust and communication bandwidth between team members. But the studied *dispersed* teams never visit and rotate with remote team members.

One of the distributed teams line managers said, *Visits to remote sites are highly costly. So, we rotate team members and mostly, the duration of the rotation between team members is 3-6 months.* - **Line Manager, Zeta project**

We have also found that teams practice *version control, screen sharing and discussion forums* to maintain knowledge sharing among both local and remote team members.

5.3   Challenges Faced by Practitioners During Knowledge Sharing Among Distributed Teams

In agile software development most of the knowledge is tacit, which resides in the human mind rather than documentation. This codified tacit knowledge is shared among between locally and globally distributed team members through tools. The knowledge sharing approach varies between team members due to experience levels. The types of problem this leads to are search availability and difficulty finding the right knowledge at the right time. We have also found that to share tacit knowledge between remote team members, teams maintain a *common chat room* and *online conference* (see in Table 3). In one project (β), we found that the team does not share tacit knowledge among dispersed team members. But, based on the situation, sometimes the team performs pair programming through screen sharing, to resolve problems. Challenges faced by the practitioners during knowledge sharing among distributed team members are shown in Fig. 1. Mitigation techniques applied by practitioners are also shown in the same Fig. 1.

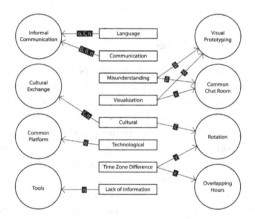

Fig. 1. Knowledge sharing challenges and mitigation techniques

In Fig. 1, arrows indicate the mitigation techniques applied by practitioners for a specific challenge. Based on the severity of *communication, language* and *cultural* challenges frequently faced by practitioners during knowledge sharing in distributed agile projects (see Fig. 1). Distanced teams are also struggling with *misunderstanding* and *visualization* challenges.

Though teams face different types of challenges during knowledge sharing among distributed team members, we identified successful knowledge sharing in

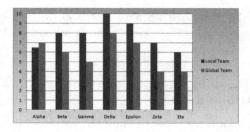

Fig. 2. Success of knowledge sharing

both locally and globally distributed agile teams from the seven cases studied. Based on the seven cases the above graph (see Fig. 2) has been drawn. An *ordinal scale* is used to map the interviewee satisfaction with their KM activities in both local and global teams. Figure 2 depicts that in project (*Alpha*) interviewees think knowledge sharing activities are more successful among distributed team members than in the local team, due to language barriers (*non-native English speaker*) between local team members. It is also evident from the gathered data that *dispersed* feature teams are more successful in their knowledge sharing among distributed team members than *distributed* and *hybrid* teams.

6   Lesson Learned

6.1   Codification of Knowledge

According to Polanyi (Polanyi 2009), *Individuals know more than they can say.* Polanyi classified human knowledge into two categories. *Tacit knowledge*, which is very difficult to describe or express: this type of knowledge is transferred through demonstration. Tacit knowledge has an important cognitive dimension which consists of mental models, beliefs and perspectives (Nonaka 1994, Bennet and Tomblin 2006; Nonaka 2007). So it cannot be easily characterized by clear expressive language. *Explicit knowledge*, is easily written down and codified. It is easily possible to characterize explicit knowledge in textual or symbolic forms. This kind of knowledge resides in textbooks, memos and technical documents. Codification of knowledge is the conversion of tacit knowledge into explicit knowledge in a written, verbal or visual format. The extraction process of tacit knowledge into explicit is called *externalization*. Tacit knowledge cannot be interpreted fully even by an expert (Ahmed et al. 2012). This type of knowledge is more deeply placed in action and is hard to express in words (Hislop 2002). Nelson *et al.* (Nelson and Winter 1982) conclude that it is impossible to describe all the necessary aspects of organizational tacit knowledge for successful performance. In organizations, most of the tacit knowledge is work related, which is learned informally as the team works (Wagner and Sternberg 1987). Codification extract tacit knowledge into explicit; it is a challenging task, so an expert needs to understand the essence of the tacit knowledge in order to increase the degree of explicitness of knowledge. Surprisingly, we found from our results that all studied cases are concerned about knowledge codification. To codify tacit knowledge, teams are using Wiki, JIRA, Confluence etc. In local sites, technical presentations and discussion forums are also taken into account as knowledge codification strategies. Later, teams share codified knowledge among remote team members through repositories and that is helpful for the remote team members to reuse codified stored knowledge.

6.2 Knowledge Management Strategies in Practices

We found that knowledge management schools are in practice, according to the results in Tables 2 and 3. We used Earls (Earl 2001) framework to select types of strategies, "schools" or practices in the different projects that applied to managing knowledge locally and globally.

Based on the evidence from the different cases, we found that knowledge management schools were in use to manage knowledge both locally and globally. It is also evident that *systems, cartographic, engineering, organizational* and *spatial* schools are practiced in distributed agile projects to manage knowledge both locally and globally. Both *commercial* and *strategic* schools are focused on a business perspective (*patent, copyright, trademark, know-how* and *intellectual assets* (Earl 2001)) and there is also no evidence found within gathered data sets that indicates those schools (*commercial* and *strategic*) are in practice. For that reason those schools are not taken into account in this research (Fig. 3).

6.2.1 Systems School

This schools philosophy is to codifying knowledge with the help of technology. Organizations use repositories for storing and sharing knowledge. These knowledge repositories usually store domain specific information. The codification of knowledge can be compared with the *externalization* of knowledge by Nonaka (Dingsøyr et al. 2009). It is easy to realize the benefits of knowledge bases and the systems school is the most researched school (Bjørnson and Dingsøyr 2008). These knowledge bases become richer and more useful over time. As shown in the Fig. 2, the *systems* school is in practice in all cases to manage knowledge locally and globally. Though search functions are a difficult issue in the systems school, practitioners depend on it because across distances this school effectively perform knowledge sharing activities using repositories.

6.2.2 Cartographic School

This school focuses on the mapping of organizational knowledge and aims to build knowledge directories by disclosing who knows what (Earl 2001). This is sometimes achieved by yellow-pages, which ensure the accessibility to others of a knowledgeable person within the organization for knowledge exchange. Though knowledge maps and directories on company intranets might be helpful for distributed team members to have an idea of who knows what, in distributed projects it seems challenging to put into practice. This is because it needs

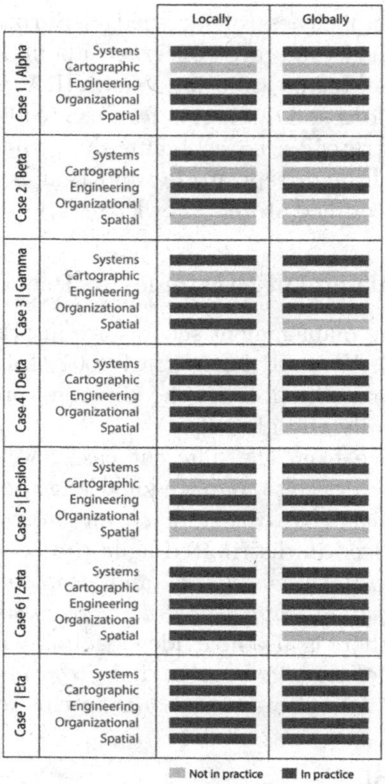

Fig. 3. Knowledge sharing strategies in practice

joint effort and commitment from both local and remote team members. In collocation, it seems easier to find a knowledgeable or experienced person because they knew each other well. In globally distributed projects who knows what and what is where are important issues for effective knowledge exchange. We have found that the *cartographic* school is practiced by different projects (δ, ζ, η) to exchange knowledge both locally and globally. This strategy is also practiced in different companies by introducing the idea of knowledge brokers: this helps other developers to consult with knowledgeable and experienced software engineers (Schneider 2009). Knowledge brokers are knowledgeable and experienced software engineers who will communicate with other developers, provide them with information or listen to them.

6.2.3   Engineering School

This school of knowledge management focuses on business process re-engineering (Bjørnson and Dingsøyr 2008) and knowledge flows in organizations. This school has a more empirical attention than other schools, which focuses on managing knowledge about software development processes and improvement of software development processes. More specifically, this school focuses on formal routines, mapping of knowledge flows, project reviews, and social interactions. Software process improvements like CMMI can be regarded as a stimulus for knowledge flow throughout the organization. This school supports explicit knowledge sharing and in distributed projects, temporal distance does not affect this school. In globally distributed projects coordination is one the major challenges and the engineering school focuses on the coordination process and aims to ensure knowledge flows within the organization though shared databases. The processes of using tools *(i.e. the installation manual for GitHub or SVN with eclipse)*, quality code writing techniques, testing and reviews are all documented in repositories to share among distributed teams. Practice of this school is found in the studied projects.

6.2.4   Organizational School

The philosophy of the *organizational* school is to create a network by collaborating between communities to share or pool knowledge. This school of knowledge management focuses on organizational structure. These structures are often referred as "knowledge communities" (Bjørnson and Dingsøyr 2008). This is a networking approach for people to communicate and share knowledge. Based on the seven cases, this school is in practice for knowledge sharing both locally and globally.

6.2.5   Spatial School

The intention of the spatial school is to encourage socialization (tacit to tacit knowledge) as a means of knowledge exchange (Lloria 2008). The spatial school is more concerned with the development and utilization of the social capital which develops from people interactions, formal or informal, repeatedly over time (Earl 2001). The spatial school focuses on designing office space to promote knowledge sharing (Dingsøyr et al. 2009). Organizations use different office settings to promote communication between people. For example, in the case of software organizations agile methodologies may use boards, charts or other tools to create spatial knowledge. Sometimes, even common spaces like conference rooms, dining rooms or places for refreshment and activities are also places where knowledge can be shared. Five cases were found that practice the *spatial* school to manage knowledge locally and only one case *(hybrid team, η)* was found that to practice the *spatial* school to manage knowledge both locally and globally. The *hybrid team* uses visual boards to communicate among collocated and distributed team members.

6.3   The Process of Organizational Knowledge Creation

Knowledge creation is continuous and dynamic process; and people are benefited from that through sharing and interacting with each other (Ernst and Kim 2002). Based on *explicit* and *tacit* knowledge, Nonaka *et al.* proposed dynamic knowledge creation model (Nonaka 1994). In this model, the authors discussed about four

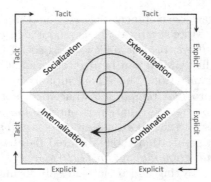

Fig. 4. Nonaka's dynamic knowledge creation model (Nonaka 1994)

types of knowledge conversion modes. The conversion mode between tacit to tacit called *socialization*, between tacit to explicit it is called *externalization*, between explicit to explicit it is called *combination* and between explicit to tacit it is called *internalization*. Each modes can independently create knowledge (Fig. 4).

6.3.1   Socialization- Tacit to Tacit

Socialization facilitate to share tacit knowledge through shared experience. In this context, the learning process start through observation, interaction and practice. So, as a result of this process technical skills and mental models are created and shared. Trust is the main ingredient in order to foster socialization (*social interaction*) in the organization (Nonaka and Takeuchi 1995). The main intension of the tacit knowledge is to learn and gain experience from others; so knowledge is created by learning during the discussion in front of coffee machine, meeting, training and team work (Nonaka 1994; Bolisani and Scarso 1999; Haldin-Herrgard 2000; Marwick 2001). In this research, we have found, team members share theirs knowledge infront of coffee machine. All studied projects share tacit knowledge among both collocated and distributed teambers.

6.3.2   Externalization- Tacit to Explicit

The extraction process of tacit knowledge into explicit is called *externalization*. In this book (Ahmed et al. 2012) the authors claimed that, tacit knowledge cannot be interpret fully even by an expert. This type of knowledge is more deeply placed in action and stiff to express in word (Hislop 2002). Nelson *et al.* (Nelson and Winter 1982) conclude, it is impossible to describe all necessary aspects of organizational tacit knowledge for successful performance. In another research Wanger *et al.* (Wagner and Sternberg 1987) mentioned, in the organization most of the tacit knowledge is work related that is learned informally during the team works. *Codification* is challenging to extract tacit knowledge into explicit; so an expert needs to understand essence of the tacit knowledge to increase degree of explicitness of knowledge. Through this study we have found, nowadays, almost all small, medium and large scale software development firms start using tools[1] to facilitate explicit knowledge sharing throughout the both collocated and distributed teams. Experience team members helping each others indirectly[2] with their experience is the distributed environment using those tools, by the problem solving, manual creation and individuals internalize what they experience.

6.3.3   Combination- Explicit to Explicit

The conversion strategy of existing explicit knowledge into new explicit knowledge is called *combination*. In the organization it is happen through meeting and conversation. The main intension of combination process is to manage all unstructured knowledge into one place by sorting, adding, categorizing and defining context. This process will help the team members to get right information at the right time. In the distributed environment team members might benefited from this process.

6.3.4   Internalization- Explicit to Tacit

Express explicit knowledge and convert it into tacit knowledge is called *internalization*. People read documents previously stored knowledge from the repositories and try to learn from that to enhance their existing knowledge. By applying this process explicit knowledge become tacit and it helps if knowledge is verbalized in documents and expressed through oral stories.

6.4   A Relation with the Concept of "Ba" (Nonaka and Konno 1998)

Japanese philosopher Kitro Nishida originally proposed this *"ba"* concept, "Ba" means *place* (Nonaka and Konno 1998). This space can be physical (i.e. office, dispersed or distributed team), virtual (i.e. email, teleconference), mental (i.e., shared experiences, idea), or any combination of them. *Ba* consider as a shared space that servers as a foundation for knowledge creation. In this research, we have found that, practitioners also apply ba through spatial school. The intention of the spatial school is to encourage socialization (tacit to tacit knowledge) as a means of knowledge exchange (Fig. 5).

6.5   Shared Understanding Between Team Members

Sucessful development and deployment depends on the shared understanding between stackholders and software engineers. According to Martin et al. [Glinz and Fricker 2014], there are two types of facts among group of people during shared understanding: explicit shared understanding (ESU) which denotes requirements, design documents, and manuals. Implicit shared understanding

---

[1] wiki's, redmine, JIRA etc.

[2] *Reading and listening to success stories make people feel the truth and root of the story (Nonaka and Takeuchi 1995).*

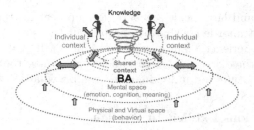

Fig. 5. Ba and knowledge conversion (Nonaka and Konno 1998)

denotes the common understanding of non-specified knowledge, such as assumptions, opinions, and values. Organization that facilitate values, structures, and practices that helps to coordinate and communicate with team members effectively. During this study, we have found, studied projects apply *"Pair programming, customer collaboration, technical presentation and so forth* to establish shared understanding among distributed team members. In addition to that, we have also observed that, team also apply both *tacit* and *explicit* knowledge sharing strategies to build shared understanding among distributed team members as in software engineering both strategies implies shared understanding.

Martin et al. [Glinz and Fricker 2013] identified the enablers and obstacles for shared understanding. Domain knowledge, Previous joint work or collaboration, Existence of reference systems, Culture and Values, Geographic distance, and Trust are enable and foster shared understanding. Contractual situation, Outsourcing, Regulatory constraints, Normal vs. radical design, Team size and diversity, and Fluctuation are an obstacle to both implicit and explicit shared understanding. Achivinging shared understanding among distributed team members is not easy though. To achieve shared understanding among distributed team members its required to overcome obstracles and enable the knowledge sharing and creation techniques to establish shared understanding.

7   Conclusion and Future Work

7.1   Summary

The aim of this research was to discover the knowledge sharing techniques, strategies applied and challenges faced by the practitioners in distributed agile projects. To perform knowledge management activities in a distributed agile project, different teams practice different types of approaches. But, in general, we found that different types of knowledge creation and sharing techniques are applied by practitioners to perform knowledge management activities in distributed agile projects. Along with that, we also found different types of strategies practiced by the team members to manage knowledge both locally and globally.

7.2   Contributions

For the first research question, we found that:

- To perform shared knowledge creation in a distributed agile project, team members practice: pair programming, customer collaboration, Scrum/Kanban boards, innovation boards, workshops/seminars, learning, technical presentation and technical discussion techniques.
- In globally distributed agile projects, teams practice different types of strategies to perform shared knowledge creation such as: *systems, engineering, organizational and cartographic* schools. We observed that the spatial school is in practice for local knowledge creation but when the project is distributed, this school is used less due to expensive tools.

For the second research question, we found that:

- To share knowledge among distributed sites, team members practice different type of techniques: repositories, pair programming, version control, screen sharing, daily scrums, weekly sprint status, common chat rooms, technical forums, discussion forums, electronic boards, online conferences, rotations/visits etc.
- *Systems, engineering* and *organizational* school strategies are explicitly in practice to share knowledge among distributed team members. These strategies foster effective knowledge sharing activities for team members in distributed agile projects. In distributed development, who knows what and what is where need to be known by employees, for effective knowledge sharing: this is associated with the cartographic school. But, in distributed agile projects, this school has is used less due to social-cultural distances. The spatial school facilitates knowledge sharing by using office space but in distributed agile projects this strategy is not explicitly in practice to share knowledge among remote team members.

For the third research question, we found that:

- During knowledge sharing among distributed team members, practitioners faced different types of challenges, such as: language, communication, misunderstanding, visualization, cultural, technological, time zone difference and lack of information.
- To mitigate those challenges, practitioners also apply different types of mitigation techniques, such as: informal communication, cultural exchange, common platform, tools, visual prototyping, common chat rooms, rotation, and overlapping hours.

7.3   Future Work

Through a series of semi-structured interviews from agile practitioners, we investigated knowledge sharing activities in distributed agile projects. Communication, coordination and collaboration are the keys to fostering knowledge sharing between team members in agile software development. However, we have seen knowledge sharing in distributed agile projects is challenging, due to factors such as communication difficulties, language barriers and cultural barriers. To mitigate those challenges and succeed in knowledge sharing within and across borders, practitioners adopt different types of techniques to manage knowledge both locally and globally. Along with these techniques, we have also noticed that, practitioners adopt different types of strategies to manage knowledge both locally and globally. Though systems, engineering and organizational schools are explicitly in practice, the spatial school has less concern with managing knowledge in distributed agile projects. With closer observation between software engineering and schools Bjørnson and Dingsøyr found that there is a heavy focus on the systems and engineering schools [Bjørnson and Dingsøyr 2008]. There are also limited number of studies focusing on the organizational school, but no studies in software engineering were found, that focus on the spatial aspect [Bjørnson and Dingsøyr 2008]. Agile software development is more related to *socialization*, which includes the spatial schools concepts of knowledge sharing strategies. There is a lot of knowledge residing in the office space and office space fosters knowledge sharing through spatial knowledge management strategy. In the future, it will be interesting to find the spatial school being practiced in distributed agile projects.

Acknowledgement. We are thankful to our dear supervisor Dr. Darja Šmite, Associate Professor and this paper also published as a Blekinge Institute of Technology's Master thesis of 1st and 3rd authors in 2013.

References

Schneider, K.: Experience and Knowledge Management in Software Engineering. Springer, Heidelberg (2009)

Richardson, I., O'Riordan, M., Casey, V., Meehan, B., Mistrik, I.: Knowledge management in the global software engineering environment. In: Fourth IEEE International Conference on Global Software Engineering, ICGSE 2009, pp. 367–369. IEEE (2009)

Sahay, S., Nicholson, B., Krishna, S.: Global IT Outsourcing: Software Development Across Borders. Cambridge University Press, Cambridge (2003)

Boden, A., Avram, G.: Bridging knowledge distribution-the role of knowledge brokers in distributed software development teams. In: ICSE Workshop on Cooperative and Human Aspects on Software Engineering, CHASE 2009, pp. 8–11. IEEE (2009)

Holz, H., Maurer, F.: Knowledge management support for distributed agile software processes. In: Henninger, S., Maurer, F. (eds.) Advances in Learning Software Organizations. LNCS, vol. 2640, pp. 60–80. Springer, Heidelberg (2003)

Šmite, D., Moe, N.B., Ågerfalk, P.: Agility Across Time and Space: Implementing Agile Methods in Global Software Projects. Springer, Heidelberg (2010)

Kavitha, R.K., Ahmed, I.: A knowledge management framework for agile software development teams. In: 2011 International Conference on Process Automation, Control and Computing (PACC), pp. 1–5. IEEE (2011)

Curtis, B., Krasner, H., Iscoe, N.: A field study of the software design process for large systems. Commun. ACM **31**(11), 1268–1287 (1988)

Nicholson, B., Sahay, S.: Embedded knowledge and offshore software development. Inf. Organ. **14**(4), 329–365 (2004)

Davenport, T.H., Prusak, L.: Working Knowledge: How Organizations Manage What They Know. Harvard Business Press, Boston (2000)

Rus, I., Lindvall, M., Sinha, S.: Knowledge management in software engineering. IEEE Softw. **19**(3), 26–38 (2002)

O'Dell, C., Grayson, C.J.: If only we knew what we know: identification and transfer of internal best practices. Calif. Manage. Rev. **40**, 154–174 (1998)

Chau, T., Maurer, F.: Knowledge sharing in agile software teams. In: Lenski, W. (ed.) Logic versus Approximation. LNCS, vol. 3075, pp. 173–183. Springer, Heidelberg (2004)

Dorairaj, S., Noble, J., Malik, P.: Knowledge management in distributed agile software development. In: Agile Conference (AGILE) 2012, pp. 64–73. IEEE (2012)

Earl, M.: Knowledge management strategies: toward a taxonomy. J. Manage. Inf. Syst. **18**(1), 215–233 (2001)

Hansen, M.T., Nohria, N., Tierney, T.: Whats your strategy for managing knowledge? Knowl. Manage. Crit. Perspect. Bus. Manage. **77**(2), 322 (2005)

Nonaka, I.: A dynamic theory of organizational knowledge creation. Organ. Sci. **5**(1), 14–37 (1994)

Dingsøyr, T., Bjørnson, F.O., Shull, F.: What do we know about knowledge management? practical implications for software engineering. IEEE Softw. **26**(3), 100–103 (2009)

Bjørnson, F.O., Dingsøyr, T.: A survey of perceptions on knowledge management schools in agile and traditional software development environments. In: Abrahamsson, P., Marchesi, M., Maurer, F. (eds.) Agile Processes in Software Engineering and Extreme Programming. LNBIP, vol. 31, pp. 94–103. Springer, Heidelberg (2009)

Nerur, S., Mahapatra, R.K., Mangalaraj, G.: Challenges of migrating to agile methodologies. Commun. ACM **48**(5), 72–78 (2005)

Nonaka, I., Konno, N.: The concept of ba: building a foundation for knowledge creation. Calif. Manage. Rev. **40**(3), 40–54 (1998)

Happel, H.-J., Maalej, W.: A lightweight approach for knowledge sharing in distributed software teams. In: Yamaguchi, T. (ed.) PAKM 2008. LNCS (LNAI), vol. 5345, pp. 14–25. Springer, Heidelberg (2008)

Hildenbrand, T., Geisser, M., Kude, T., Bruch, D., Acker, T.: Agile methodologies for distributed collaborative development of enterprise applications. In: International Conference on Complex, Intelligent and Software Intensive Systems, CISIS 2008, pp. 540–545. IEEE (2008)

Boden, A., Avram, G., Bannon, L., Wulf, V.: Knowledge management in distributed software development teams-does culture matter? In: Fourth IEEE International Conference on Global Software Engineering, ICGSE 2009, pp. 18–27. IEEE (2009)

Marczyk, G.R., DeMatteo, D., Festinger, D.: Essentials of Research Design and Methodology, vol. 2. Wiley, New York (2010)

Runeson, P., Höst, M.: Guidelines for conducting and reporting case study research in software engineering. Empirical Softw. Eng. **14**(2), 131–164 (2009)

Sekaran, U.: Research Methods for Business: A Skill Building Approach. Wiley, New York (2006)

Creswell, J.W.: Research Design: Qualitative, Quantitative, and Mixed Methods Approaches. Sage Publications, Incorporated, Thousand Oaks (2008)

Flick, U.: An Introduction to Qualitative Research. Sage Publications Limited, London (2009)

Robson, C.: Real World Research: A Resource for Social Scientists and Practitioner-researchers, vol. 2. Blackwell Oxford, Malden (2002)

Basit, T.: Manual or electronic? the role of coding in qualitative data analysis. Educ. Res. **45**(2), 143–154 (2003)

Braun, V., Clarke, V.: Using thematic analysis in psychology. Qual. Res. Psychol. **3**(2), 77–101 (2006)

Guest, G., MacQueen, K.M., Namey, E.E.: Applied Thematic Analysis. Sage Publications, Incorporated, Thousand Oaks (2011)

Miles, M.B., Huberman, A.M.: Qualitative Data Analysis: An Expanded Sourcebook. Sage Publications, Incorporated, Thousand Oaks (1994)

Punch, K.F.: Introduction to Research Methods in Education. Sage Publications Limited, Thousand Oaks (2009)

Shull, F., Singer, J., Sjøberg, D.I.K.: Guide to Advanced Empirical Software Engineering. Springer, London (2007)

Bleijenbergh, I., Korzilius, H., Verschuren, P.: Methodological criteria for the internal validity and utility of practice oriented research. Qual. Quant. **45**(1), 145–156 (2011)

Cress, U., Kimmerle, J.: A systemic and cognitive view on collaborative knowledge building with wikis. Int. J. Comput. Support. Collaborative Learn. **3**(2), 105–122 (2008)

Palmieri, D.W.: Knowledge management through pair programming (2002)

McDermott, R.: Learning across teams. Knowl. Manage. Rev. **8**(3), 32–36 (1999)

Sureshchandra, K., Shrinivasavadhani, J.: Adopting agile in distributed development. In: IEEE International Conference on Global Software Engineering, ICGSE 2008, pp. 217–221. IEEE (2008)

Moe, N.B., Šmite, D.: Understanding a lack of trust in global software teams: a multiple-case study. Softw. Process: Improv. Pract. **13**(3), 217–231 (2008)

Polanyi, M.: The Tacit Dimension. University of Chicago Press, Chicago (2009)

Bennet, A., Tomblin, M.S.: A learning network framework for modern organizations: organizational learning, knowledge management and ict support. VINE **36**(3), 289–303 (2006)

Nonaka, I.: The knowledge-creating company. Harvard Bus. Rev. **26**(4–5), 598–600 (2007)

Ahmed, P.K.K., Lim, K.K.K., Loh, A.Y.: Learning Through Knowledge Management. Routledge, London (2012)

Hislop, D.: Mission impossible? communicating and sharing knowledge via information technology. J. Inf. Technol. **17**(3), 165–177 (2002)

Nelson, R.R., Winter, S.G.: An Evolutionary Theory of Economic Change. Belknap Press, Cambridge (1982)

Wagner, R.K., Sternberg, R.J.: Tacit knowledge in managerial success. J. Bus. Psychol. **1**(4), 301–312 (1987)

Bjørnson, F.O., Dingsøyr, T.: Knowledge management in software engineering: a systematic review of studied concepts, findings and research methods used. Inf. Softw. Technol. **50**(11), 1055–1068 (2008)

Lloria, M.B.: A review of the main approaches to knowledge management. Knowl. Manage. Res. Pract. **6**(1), 77–89 (2008)

Ernst, D., Linsu, K.: Global production networks, knowledge diffusion, and local capability formation. Res. Policy **31**(8), 1417–1429 (2002)

Nonaka, I., Takeuchi, H.: The Knowledge-Creating Company: How Japanese Companies Create the Dynamics of Innovation. Oxford University Press, USA (1995)

Bolisani, E., Scarso, E.: Information technology management: a knowledge-based perspective. Technovation **19**(4), 209–217 (1999)

Haldin-Herrgard, T.: Difficulties in diffusion of tacit knowledge in organizations. J. Intellect. Capital **1**(4), 357–365 (2000)

Marwick, A.D.: Knowledge management technology. IBM Sys. J. **40**(4), 814–830 (2001)

Glinz, M., Fricker, S.A.: On shared understanding in software engineering: an essay. Computer Science-Research and Development, pp. 1–14 (2014)

Glinz, M., Fricker, S.: On shared understanding in software engineering. In: Software Engineering, pp. 19–35. Citeseer (2013)

Actuator Fault Diagnosis Using Single and Meta-Classification Strategies

Mateusz Kalisch[✉], Piotr Przystałka, and Anna Timofiejczuk

Institute of Fundamentals of Machinery Design, Silesian University of Technology,
18a Konarskiego Street, 44-100 Gliwice, Poland
mateusz.kalisch@polsl.pl

Abstract. The paper presents the application of various classification schemes for actuator fault diagnosis in industrial systems. The main objective of this study is to compare either single or meta-classification strategies that can be successfully used as reasoning means in the diagnostic expert system that is realized within the frame of the DISESOR project. The applied research was conducted on the assumption that classic as well as soft computing classification methods would be adopted. The comparison study was carried out within the DAMADICS benchmark problem which provides a popular framework for confronting different approaches in the development of fault diagnosis systems.

1 Introduction

The increasing complexity of recent industrial objects makes the issue of fault diagnosis one of the most important directions of research in modern automatic control and robotics [7,24,32]. Technical systems and processes are required to be safely and reliably operated due to the protection of human life and health, the quality of the environment, as well as the economic interests. It is possible to specify numerous areas of interdependence of human and technical means, where safety plays a key role, for instance in aircraft, spaceship, automotive, power or mining industry. The above mentioned factors cause that new developments in control theory such as passive and active fault-tolerant control approaches are more often applied in these areas of the industry [5,17,22]. A special attention is currently paid on the second type of the advanced control methodologies, where fault diagnosis methods hold a critical importance. The present state of the art in the field of fault diagnosis shows the really need for development of fault diagnosis expert systems. The goal is to elaborate general-purposes systems with multi-domain knowledge representations and multi-inference engines [9,28,36]. Generally, the fault diagnosis can be divided into three steps [18]: fault detection, fault isolation and fault identification. Moreover, each of them can be developed by means of model-free (based on data), model-based and knowledge-based approaches [22]. In this paper the first approach, where experimental data are exploited was discussed. In this kind of methods data that represents normal and faulty situations can be obtained from historical databases or from simulators as

© IFIP International Federation for Information Processing 2015
E. Mercier-Laurent et al. (Eds.): AI4KM 2014, IFIP AICT 469, pp. 132–149, 2015.
DOI: 10.1007/978-3-319-28868-0_8

well as laboratory stands. This data is then used to create state classifiers and meta-classifiers.

The main goal of this paper is to compare different classification strategies that can be successfully used as reasoning means in the diagnostic expert system. The development of the diagnostic expert system shell with multi-domain knowledge representations and multi-inference engines is realized within the frame of the DISESOR project. The DISESOR is an acronym of the decision support system designed for fault diagnosis of machinery and other equipment operating in underground mines as well as for monitoring potential threats that can occur in such kind of industry. The DISESOR system can be used for different purposes, e.g. to assess seismic hazard probabilities in the area of the coal mine, to forecast dangerous increase in the methane concentration in the mine shafts, to detect and localize endogenous fires, and also to conduct fault diagnostics of machines working in such environment. This study shows the comparison research of the classification schemes for creating fault diagnosis system of the benchmark actuator [2] which was elaborated on the basis of the activity of the DAMADICS (Development and Application of Methods for Actuator Diagnosis in Industrial Control Systems) Research Training Network funded by the European Commission. The current paper is a continuation of the research work presented in [20]. The authors taken into account the majority of reviewers' comments and also proposed a new approach for searching proper values of relevant parameters of classifiers used to fault diagnosis. The examined methods are planed to be used for designing the engine of the DISESOR system.

2 Single and Meta-Classification Strategies

There are many types of classifiers available in the literature, as well as different concepts of using them are introduced [25]. Some examples are methods based on the similarity between objects in the feature space, probabilistic methods or methods which are based on black box models. Generally, the classification problems can be divided into two groups including approaches of supervised and unsupervised machine learning techniques. In the paper, the authors concentrated the attention only on methods belonging to the first group. Currently, the information fusion and meta-classification problems are recognized as the most important directions of the research in the domain of supervised learning. The main idea in this approach is the application of simple classifiers working together to solve a problem with better results than it can be done by means of single one or more complicated classifiers. There are a lot of different kinds of information fusion methods, but the most popular are majority voting, weighted voting, boosting, and AdaBoost [25]. On the other hand, meta-classifiers are very often used for the same reason that means its efficiency is often higher, than the efficiency of the best single classifier [26].

The current research trends in developing machine learning methods are focused on ideas of improving the general efficiency of different classification and meta-classification methods. The most important investigations can be found for instance in [4,12,30,33,38,39]. The main directions presented in these studies are concentrated on optimization techniques which are used to tune relevant parameters of the classical methods, e.g. with the use of evolutionary and particle swarm algorithms. A number of results included in the related works show the benefits of using these methods. In case of a task of fault detection and isolation the key features of the signals in time or frequency domains are most commonly used. Industrial actuators may be characterized by a very high complexity which affects the large number of measuring signals and their features. Therefore, another approach aimed at improving the efficiency of the classifier, and often shortening the time of its learning, is to remove irrelevant variables [14]. There are various methods that can be used in this procedure, e.g. forward or backward selection methods, as well as elimination methods based on statistical measures. Another group of methods stands fusion methods such as bagging, boosting, and the development of these concepts that is AdaBoost method [19,40]. These methods are often more effective than simple classifiers but also show some drawbacks. The advanced concepts were developed to take merits and positive aspects of classic methods and to eliminate their limitations [37]. There are also attempts to connect together several different methods such as selection of relevant features and usage of boosting into one algorithm [21]. Such approach may lead to the final result that should be better than the results of the methods applied separately.

3   Model-Free Fault Diagnosis Using Different Classification Schemes

The idea of the well practised model-free fault detection and isolation method is presented in Fig. 1. It can be seen, that faults are detected and distinguished using primary and redundant process variables. In this method two separated classifiers must be created. The first classifier uses the subset of process variables $(U' \cup Y')$ as its input and it is dedicated for generating diagnostic signals (S), whereas the second one has the same set of input variables but its task is to calculate a fault signature (F). This classifier is triggered in case when the diagnostic signal indicates a fault scenario. The proposed method can be viewed as the extension of the most often used model-free fault diagnosis approach, cf. Korbicz et al. (2004), Fig. 1.7 at p. 22 [22]. The novelty in this study depends on that single and also meta-classifiers are automatically tuned in order to obtain the maximum accuracy of fault diagnosis.

Fault detection and isolation algorithms corresponding to the diagram presented in Fig. 1 can be designed using different classification methods [18,22,31]. Generally, it is possible to apply so-called classical (e.g. decision trees, k-nearest neighbour, naive Bayes, etc.) or soft computing approaches (e.g. neural networks, bayesian networks, fuzzy systems, neuro-fuzzy systems, etc.). The paper deals with either classic or soft computing methods. In the next part of the article,

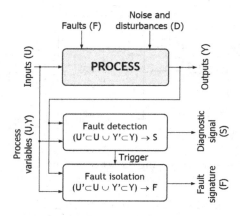

Fig. 1. A diagram of model-free fault detection and isolation

model-free fault detection and isolation approaches with the use of different classification schemes are described. As it was mentioned above, these kinds of methods require data (process variables) corresponding to regular (faultless) and faulty states of the system. In this section, different variants of three basic concepts with a single classifier, meta-classifier and a bank of classifiers are applied in order to provide the fault detection and isolation system that is directly based on the process variables.

3.1 Fault Detection Schemas

The first concept of fault detection is presented in Fig. 2 and this is elaborated basing on a single classifier which returns a diagnostic signal corresponding to fault or faultless states of the device. In this method, the process variables are converted by a moving window in order to compute scalar features of the measuring signals. These values are used as input of a single classifier which generates directly the diagnostic signal. The second fault detection scheme is presented in Fig. 3. In this approach a series of two-state classifiers is applied and their task is to determine the degree of the belief for fault detection. The level of belief about faults occurring is a numerical value from 0 to 1. The signal values returned from each classifier are connected to the meta-classifier as its input. The features of the process variables are also connected to the meta-classifier as the additional input.

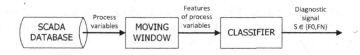

Fig. 2. A scheme of fault detection using the global classifier

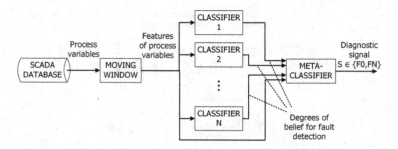

Fig. 3. A scheme of fault detection using the set of various classifiers and meta-classifier

The result of both methods is a diagnostic signal which indicates fault occurrence. When a classifier or a meta-classifier detects a fault, the second part of the fault diagnosis system is run in order to isolate the faults.

3.2 Fault Isolation Schemes

The first method of fault isolation is comparable to the method that was proposed for the fault detection. It is presented in Fig. 4. As one can see it is a single global classifier. Its task is to determine a type of the fault. Similarly to the previous method, in this case the process variables are calculated in the moving window to obtain scalar features of the measuring signals. The preprocessed signals are connected to the input of a global classifier. This classifier returns a fault signature.

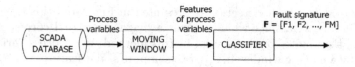

Fig. 4. A scheme of fault isolation using the global classifier

The next fault isolation scheme is presented in Fig. 5. In this approach a set of classifiers of different types is used in order to calculate the degrees of beliefs that are related to fault signatures. These values are given to the input of the meta-classifier and the final decision (fault signature) is obtained.

The last concept of fault isolation is shown in Fig. 6. The main idea is based on a bank of classifiers that are used to calculate degrees of beliefs for specific faults and unknown states of a device. In this case, M single classifiers must be created for M faulty states. Each classifier is dedicated for one state only (it is used for detection one fault solely). In the next step, all available variables (features of the process variables and outputs from base classifiers) are linked to a single dataset. The prepared signals are sent to the input of the meta-classifier which is employed to return the final decision.

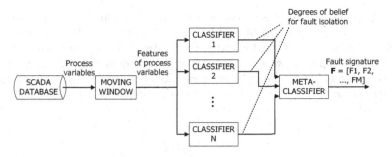

Fig. 5. A scheme of fault isolation using the set of different classifiers and meta-classifier

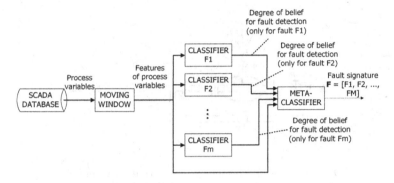

Fig. 6. A scheme of fault isolation using the set of local classifiers (fault detectors) and meta-classifier

The engines of fault detection and isolation schemes presented above can be elaborated with the use of well practised classification methods. The classification problem is possible to be solved using many known approaches, however, in this research the following methods are applied: k-nearest neighbour [1], naive Bayes [10], decision tree [6,27], rules induction [11], neural networks [13,15] and support vector machine [16]. Each of these classifiers returns a label of a chosen class and the degrees of belief for all predicted classes. The best solution is pointed at the moment when one of the class is characterised by the belief level equal to 1 and the rest of them are equal to 0. It gives us 100 % certainty that a new element should be classified as this particular class.

4   Verification Studies

The proposed schemes of fault detection and isolation were implemented using RapidMiner®software. It is an open source software created for solving data mining problems. The verification studies were conducted on data generated using the DAMADICS simulator [3] in order to investigate selected classification schemes. This simulator was elaborated in collaboration of scientists and engineers to simplify the process of evaluating and comparing different methods of fault detection and isolation for industrial systems. In the literature there are available several papers where case study results deal with this problem are presented, see e.g. [23, 29, 35]. The numeric model is used to simulate an electro-pneumatic valve (Fig. 7) which is a part of the production line in Lublin sugar factory in Poland.

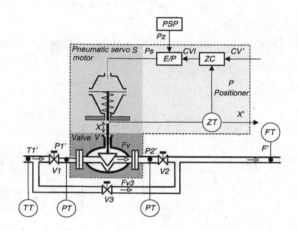

Fig. 7. Structure of benchmark actuator system [2]

The presented model was created and tuned in MATLAB/Simulink®software taking into account the physical phenomena related to the origin of faults in the real actuator system. This simulator was used to generate the following signals of the process variables: CV - process control external signal, $P1$ - inlet pressures on valve, $P2$ - outlet pressures on valve, X - valve plug displacement, F - main pipeline flow rate, T - liquid temperature, f - fault indicator. All of these signals were normalized to the range between 0 and 1.

The DAMADICS simulator allows to choose only one from nineteen available faults - a part of them is considered only as incipient faults or as abrupt faults (there are three sizes of abrupt faults: small, medium and big) and some of them as both. In this paper the authors decided to investigate only abrupt faults, such as: f1 - valve clogging, f2 - valve or valve seat sedimentation, f7 - medium evaporation or critical flow, f8 - twisted servo-motor stem, f10 - servomotor diaphragm perforation, f11 - servomotor spring fault, f12 - electro-pneumatic transducer fault, f13 - stem displacement sensor fault, f14 - pressure sensor fault, f15 - positioner spring fault, f16 - positioner supply pressure drop, f17 - unexpected pressure change across valve, f18 - fully or partly opened bypass valves, f19 - flow rate sensor fault. Moreover, only scenarios with single faults were taken into account. The list does not include some faults, because the incipient faults such as f3 - *Valve or valve seat erosion* or f4 - *Increase of valve friction* were not considered. The verification tests were performed basing on the process variables generated by the DAMADICS simulator for fault-free and faulty scenarios.

4.1 Data Preparation

Collections of data for the training, test and verification of classifiers were prepared in such a way that the results of classifiers were very similar to classifiers working in a real environment. The process of learning (training and testing) and verification for the applied classifiers was described in this section. Data preparation is a very important part of the classifier learning process. The dataset should be divided into two et last equal parts, where the first part describes correctly working device (fault-free state) and the second part corresponds to situation when fault occurs. It was important to divide the prepared data again into two separated groups (learning group and verification group). For the meta-classifier the number of groups was extended to four, because the first and the second group were used in the learning and verification process for the base classifier. The other two groups were used for a meta-classifier. In this approach, the size of the dataset for each group was equal to 25200 samples, where 12600 samples were prepared from data without faults and rest of them contained data with all considered faults.

The data prepared for the first two fault isolation methods consists of characteristic process values for all chosen faults. The number of elements for each fault was the same for all sets. For learning and verification process four independent groups of data were prepared (like in fault detection methods, two for base classifiers and two for meta-classifier). The dataset for a single fault for one group contains 900 samples, while the full dataset size is equal to 12600 samples. The third method of fault isolation requires a different type of data. The initial classifier needs data, where a half of the elements describes an actuator device working with one specific fault and the rest of the elements describe the device working with the other faults. In this case, a classifier can generate a two-state signal where the first state defines one specific fault and the other ones are correlated with unknown faults. The size of the dataset in this approach is similar as in case for the method of fault detection. The size of the dataset for the considered fault is equal to 3900 samples and it is equal to the rest of a dataset which contains samples corresponding to other faults.

4.2 Statistical Analysis and Features Computation

Linear correlation and mutual information analysis were used for choosing relevant process variables and a proper value of the width of a moving window function. In the analysis, all of available process variables were compared for different device status (e.g. device without faults and with a chosen fault). The results of these tests showed very strong correlations between states F8, F14 and F0. A group of useful process signals was prepared on the basis of results of these tests. Most of the process signals had very difficult character for model-free fault detection and isolation methods. Therefore, the authors decided to apply scalar features of the process variables. The authors employed a few well known features often used in fault detection and isolation processes such as average, maximum and minimum values, standard deviation, root mean square, shape factor (a factor determines the shape of the periodic signal), kurtosis (the measure for evaluating the shape of the probability distribution), energy, skewness (the measure of the asymmetry of the probability distribution) and entropy. The scalar features were computed using a moving window of 100 samples width. Such the width value was assumed on the basis of frequency of the harmonic control signal of the valve which was equal to $0,01$ Hz. The authors also studied other values of the window width, however, expected effects in increasing of the efficiencies of the models were not observed.

4.3   Results of Verification Studies

The learning process for the whole set of applied classifiers was conducted using the X-validation method. To show the efficiency of a classification process, the authors used three measures corresponding to accuracy, sensitivity and precision. The accuracy presents the proportion of correct guesses which can be used directly as efficiency of classification because all datasets were very good balanced (the number of samples for each class was equal). The sensitivity corresponds to the proportion of samples in actual class which are correctly identified. The precision denotes the probability that an example assigned to specific class should be really connected with this class. The sensitivity and precision were calculated for each class, whereas the accuracy were calculated as one value for the whole confusion matrix. In order to have study results more clearly in each table within this section the following notation was assumed: kNN - k-nearest neighbours, NB - Naive Bayes, DT - Decision Tree, RI - Rule Induction, NN - Neural Net, SVM - Support Vector Machine. The prefix letter M placed before each label of the classifier means the meta-version of this classifier, for example, the label MNB denotes the meta-classifier which is based on a naive Bayes classifier.

Moreover, the authors decided to apply the parameter optimization operator included in Rapidminer software in order to tune behavioural parameters of the investigated classifiers. The described schemes of fault detection and isolation were examined taking into consideration all types of classifiers. The optimization operator for each variant of a classifier was based on the evolutionary computation. The standard evolutionary algorithm was used [13]. The maximal number of generations was set to 30 and the population size was equal to 5. The tournament method was used to select parents for creating new generations (the tournament fraction was equal to 0.25). Reproduction was realized based on Gaussian mutation function and crossover operator. The crossover probability was equal to 0.9. The optimization operator maximized accuracy value of chosen classifier. In the case of Naive Bayes classifier the optimize selection operator was only applied to choose the most relevant attributes. The features of this algorithm were similar to the previous one. The k-nearest neighbours classifier was optimized by searching a proper value of the parameter k. In case of decision tree following values of parameters were chosen: the minimal leaf size, the maximal depth of tree, the minimal gain which affects the ability of the splitting of leaf as well as the confidence used for the pessimistic error calculation of pruning. The rule induction was optimized in the context of the sample ratio that specifies the sample ratio of training data used for growing and pruning, the pureness corresponds to the desired pureness and the minimal prune benefit. In the case of a neural net classifier the number of layers and the number of neurons in each layer were tested and chosen by the authors based on heuristics known from the literature. The other parameters of neural classifiers were optimized by the operator: the number of training cycles, the learning rate determines how much the weights are adjusted at each step, the momentum coefficient smooths optimization directions and the error epsilon as training error threshold. The SVM based classifier has many behavioural parameters and therefore the authors decided to use the methodology described in [16] and focused their attention only on γ and C parameters of the radial base function. The more detailed description of relevant parameters of classifiers and the optimization operator can be found in [1]. The minimum and maximum values of the optimized parameters were defined on the basis of the information contained in the literature.

Results of Fault Detection. In the first concept of fault detection (Fig. 2) six single classifiers were compared. Table 1 shows the accuracy and sensitivity of two concepts of fault detection realized relating to the schemes presented in Figs. 2 and 3. The sensitivity values of classifiers are given in columns. The column indicated as "All" includes the general efficiency calculated on the basis of the confusion matrix which was generated after the classifier verification process. The next column (F0) includes the sensitivity obtained for faultless states. The rest of the columns (F1–F19) show the sensitivity of fault detection for all considered faults, separately. Above these columns the general result of the sensitivity of fault detection is presented. Rows from 1 to 6 show results for single classifiers, whereas the next six rows show results for considered meta-classifiers. The second table (Table 2) shows the precision of fault detection for each classifier and class (faultless state and failure state). All precision values are less than 1, 00 which means that there always exist some samples which are assigned to another state than it should.

In first fault detection scheme the accuracy of most used classifiers is high (above 0, 93). The accuracy of neural net and naive bayes based classifiers is a little bit lower than other ones but it is still in the acceptable range. The sensitivity of faultless state detection is very close to 1, 00, which means that almost all samples corresponding to faultless state were classified correctly to this class. It is easy to distinguish faults which are low-correlated with faultless state because the sensitivity in this case is equal to 1, 00. Faults highly-correlated with faultless state have very low sensitivity value, sometimes even equal to zero. The precision values presented in the Table 2 shows proportion between the number of samples which belong to this class and the number of all samples assigned by classifier to the specific class. In the second scheme of fault detection meta-classifiers were used. All six classifiers trained in the previous process based on the first scheme

Table 1. The accuracy and sensitivity of fault detection for global classifiers and meta-classifiers

	All	F0	F1	F2	F7	F8	F10	F11	F12	F13	F14	F15	F16	F17	F18	F19
kNN	0,933	0,978	0,887													
			0,93	1,00	1,00	0,67	0,84	0,84	0,67	1,00	0,67	1,00	0,78	1,00	1,00	0,99
NB	0,878	0,992	0,764													
			1,00	1,00	1,00	0,00	0,65	1,00	0,71	1,00	0,00	1,00	0,33	1,00	1,00	1,00
DT	0,960	0,976	0,946													
			1,00	1,00	1,00	0,65	0,98	1,00	1,00	1,00	0,67	1,00	0,94	1,00	1,00	1,00
RI	0,957	0,973	0,939													
			1,00	1,00	1,00	0,67	0,95	1,00	0,98	1,00	0,72	1,00	0,81	1,00	1,00	1,00
NN	0,829	0,977	0,681													
			0,75	1,00	1,00	0,24	0,49	0,17	0,25	1,00	0,24	1,00	0,50	1,00	1,00	0,89
SVM	0,938	0,982	0,894													
			0,96	1,00	1,00	0,67	0,86	0,95	0,67	1,00	0,67	1,00	0,78	1,00	1,00	0,96
MkNN	0,862	0,931	0,793													
			1,00	1,00	1,00	0,01	0,84	1,00	0,92	1,00	0,01	1,00	0,34	1,00	1,00	1,00
MNB	0,889	0,980	0,799													
			1,00	1,00	1,00	0,00	0,87	1,00	0,99	1,00	0,00	1,00	0,33	1,00	1,00	1,00
MDT	0,883	0,971	0,794													
			1,00	1,00	1,00	0,00	0,83	1,00	0,95	1,00	0,00	1,00	0,33	1,00	1,00	1,00
MRI	0,888	0,974	0,803													
			1,00	1,00	1,00	0,00	0,85	1,00	1,00	1,00	0,00	1,00	0,39	1,00	1,00	1,00
MNN	0,875	0,926	0,824													
			1,00	1,00	1,00	0,15	0,86	1,00	0,93	1,00	0,15	1,00	0,45	1,00	1,00	1,00
MSVM	0,773	0.959	0.587													
			0,69	0,89	0,75	0,00	0,57	0,87	0,63	0,39	0,00	0,88	0,23	0,87	0,69	0,76

(Fig. 3) were used to calculate degrees of belief of fault detection. The obtained data were merged with base classifier's input data. The final results of classification for the second scheme of fault detection is presented in Tables 1 and 2. In most cases (except classifier based on neural net and naive Bayes) the general accuracy, sensitivity and precision of classification in the second scheme are lower or similar to analogous measures in the first scheme. The meta-classifier based on neural net provided the higher accuracy, sensitivity and precision than neural net used as a simple classifier. It means that additional variables (degrees of beliefs of base classifiers) can be successfully applied to improve the efficiency of classification.

Table 2. The precision of fault detection for global classifiers and meta-classifiers

	kNN	NB	DT	RI	NN	SVM	MkNN	MNB	MDT	MRI	MNN	MSVM
F0	0,90	0,81	0,95	0,94	0,75	0,90	0,82	0,83	0,83	0,83	0,84	0,70
F1–F19	0,98	0,99	0,97	0,97	0,97	0,98	0,92	0,98	0,97	0,97	0,92	0,93

Results of Fault Isolation. The last three methods concerned fault isolation without taking into account a faultless scenario. The results of comparison of the first fault isolation method (Fig. 4) are included in Tables 3 and 4. The column indicated as "All" in Table 3 includes values of the accuracy for single classifiers of different types. The rest of the columns show information deals with the sensitivity of fault isolation for each scenario.

Table 3. Accuracy and sensitivity of fault isolation for global classifiers

	All	F1	F2	F7	F8	F10	F11	F12	F13	F14	F15	F16	F17	F18	F19
kNN	0,73	0,84	1,00	1,00	0,35	0,76	0,72	0,51	0,81	0,05	1,00	0,48	1,00	1,00	0,67
NB	0,77	0.67	1,00	1,00	0,58	0,63	1,00	1,00	0,58	0,40	1,00	0,28	1,00	0,91	0,67
DT	0,07	1,00	0,00	0,00	0,00	0,00	0,00	0,00	0,00	0,00	0,00	0,00	0,00	0,00	0,00
RI	0,78	1,00	1,00	1,00	0,59	0,67	1,00	0,98	0,61	0,00	1,00	0,73	1,00	0,63	0,67
NN	0,77	1,00	1,00	1,00	0,00	0,62	0,96	1,00	0,81	0,00	1,00	0,65	1,00	1,00	0,67
SVM	0,67	0,48	0,96	1,00	0,09	0,78	0,95	0,79	0,64	0,00	0,98	0,50	0,91	0,63	0,67

Table 4 presents the precision of classification for each class and classifier. The results of tested classifiers (Table 3) were varied, e.g. the accuracy of decision tree was 0.07 and the second one (sorted by their accuracy values) was 0.67. The sensitivity of decision tree for fault 1 is equal to $1, 00$ but the precision is equal to $0, 07$. It means that a classifier is able to classify correctly samples related to fault 1 but unfortunately the rest of samples connected with other faults are also classified as fault 1. This kind of classifier is not useful. The accuracy of the next group of classifiers are better and rule induction based ones reach the best result equal to $0, 78$. In the sensitivity table (Table 2) all classifiers (except decision tree) reach very similar results for most of the classes. The sensitivity value is more varied for classes which are not easily distinguishable e.g. classification sensitivity for fault 16 is the range from $0, 00$ to $0, 73$ and each classifier reach a different value. The precision values (Table 4) are correlated with the sensitivity value. For faults which are easy to classify the precision is equal or close to $1, 00$, for more difficult faults to recognize the precision values are varied.

Table 4. The precision of fault isolation for global classifiers

	F1	F2	F7	F8	F10	F11	F12	F13	F14	F15	F16	F17	F18	F19
kNN	0,99	1,00	1,00	0,27	0,61	0,50	0,47	0,99	0,15	1,00	0,61	0,75	0,84	1,00
NB	1,00	1,00	1,00	0,34	0,69	0,86	0,97	0,87	0,35	1,00	0,73	0,75	0,68	1,00
DT	0,07	0,00	0,00	0,00	0,00	0,00	0,00	0,00	0,00	0,00	0,00	0,00	0,00	0,00
RI	0,75	1,00	1,00	0,47	1,00	0,88	0,95	0,58	0,00	1,00	0,37	1,00	0,74	1,00
NN	1,00	1,00	1,00	0,05	0,94	1,00	0,57	0,71	0,00	1,00	0,26	1,00	0,84	0,99
SVM	0,99	1,00	0,34	0,39	0,72	0,92	0,72	0,72	0,04	1,00	0,26	1,00	0,99	0,95

The second method presented in Fig. 5 uses six classifiers as in the previous method but the outputs of these classifiers are connected to a meta-classifier. The results obtained for the meta-classifier are compared in Tables 5 and 6. The results obtained for the second scheme of fault isolation (Fig. 5), which was based on the meta-classifier were more similar to each other than in the first case (Fig. 4). The accuracy of classification (Table 5) in the second scheme is much better than in the first approach of fault isolation. Meta-classifiers used only degrees of beliefs from base classifiers and they were able to significantly increase the general efficiency. The most of precision values (Table 5) for each class are very similar. The similarity between sensitivity and precision tables (Table 6) is very high but there is one exception. The sensitivity of fault 12 is less then 1, 00 but the precision for almost all classifiers is equal to 1, 00. It means that not all samples connected with fault 12 are recognized but one can be sure that all recognized samples are really connected with this fault.

The last method of fault isolation (Fig. 6) is based on series of single classifiers, where each classifier is used for detecting a single fault. The first task of the verification process was to choose a single classifier (from six available) for the fault detection purpose. To solve this problem the authors tested all classifiers for all available faults. The results are presented in Table 7. The values included in the table present the general accuracy of each classifier. The bold values are related to the classifiers which were chosen as the basic classifiers for the meta-classifier.

Table 5. The accuracy and sensitivity of fault isolation for meta-classifiers

	All	F1	F2	F7	F8	F10	F11	F12	F13	F14	F15	F16	F17	F18	F19
MkNN	0,86	1,00	1,00	1,00	0,63	0,98	0,99	0,88	0,98	0,11	1,00	0,53	1,00	1,00	1,00
MNB	0,84	0,97	1,00	1,00	0,14	0,91	0,99	0,78	0,97	0,65	1,00	0,49	1,00	0,91	1,00
MDT	0,84	0,99	1,00	1,00	0,00	0,84	0,99	0,74	0,97	0,84	1,00	0,43	1,00	1,00	1,00
MRI	0,85	0,99	1,00	1,00	0,70	0,99	1,00	0,74	0,98	0,14	1,00	0,40	1,00	1,00	1,00
MNN	0,87	1,00	1,00	1,00	1,00	0,95	1,00	0,88	0,97	0,00	1,00	0,41	1,00	1,00	1,00
MSVM	0,78	0,94	1,00	1,00	0,00	0,89	0,99	0,81	0,93	0,00	1,00	0,99	1,00	0,38	1,00

Table 6. The precision of fault isolation for meta-classifiers

	F1	F2	F7	F8	F10	F11	F12	F13	F14	F15	F16	F17	F18	F19
MkNN	1,00	1,00	1,00	0,40	0,94	0,93	0,74	1,00	0,33	1,00	0,66	1,00	0,98	1,00
MNB	0,90	1,00	1,00	0,36	0,98	1,00	1,00	0,91	0,38	1,00	0,44	1,00	0,97	1,00
MDT	0,99	1,00	1,00	0,00	1,00	0,86	1,00	1,00	0,37	1,00	0,46	0,99	0,97	0,99
MRI	1,00	1,00	1,00	0,39	0,68	0,99	1,00	1,00	0,28	1,00	0,79	0,99	0,98	1,00
MNN	1,00	0,99	1,00	0,38	0,99	0,90	1,00	1,00	0,00	1,00	0,99	1,00	0,97	1,00
MSVM	1,00	1,00	1,00	0,00	0,99	1,00	1,00	1,00	0,00	1,00	0,25	1,00	0,95	1,00

Table 7. The comparison results of base classifiers for fault isolation of single faults

	F1	F2	F7	F8	F10	F11	F12	F13	F14	F15	F16	F17	F18	F19
kNN	0,92	1,00	1,00	0,72	0,80	0,84	0,80	**0,98**	0,72	1,00	0,74	1,00	**0,98**	0,98
NB	0,64	**1,00**	1,00	0,92	0,71	0,99	0,89	0,94	**0,90**	1,00	0,72	1,00	0,96	**1,00**
DT	**1,00**	1,00	1,00	0,77	0,59	**1,00**	**0,95**	0,92	0,79	1,00	0,63	0,99	0,91	0,81
RI	0,93	1,00	1,00	**0,87**	0,62	1,00	0,92	0,94	0,85	1,00	0,56	0,99	0,94	0,89
NN	0,94	0,99	**1,00**	0,81	0,85	0,99	0,89	0,97	0,81	1,00	**0,80**	1,00	0,98	0,71
SVM	0,97	0,99	0,98	0,79	**0,88**	0,92	0,86	0,92	0,99	0,80	0,80	0,99	0,96	0,97

In the next step of the method the meta-classifier is used. Its inputs are connected to the outputs of basic classifiers (the degrees of the belief for single fault detection). The main task of this meta-classifier is to compute the final result. Table 8 presents the sensitivity of different types of classifiers which are presented in the same form as in the second method of fault isolation (Table 5). In the first column indicated as "All" there are included values of the accuracy of the meta-classifiers. In the next columns the sensitivity values of single fault isolation obtained by means of meta-classifiers are included.

Table 8. The accuracy and sensitivity of fault isolation for meta-classifiers with a bank of classifiers for isolating single faults

	All	F1	F2	F7	F8	F10	F11	F12	F13	F14	F15	F16	F17	F18	F19
MkNN	0,80	1,00	1,00	1,00	0,28	0,74	0,86	0,94	0,81	0,21	1,00	0,42	1,00	0,99	0,98
MNB	0,76	1,00	1,00	1,00	0,46	0,55	0,93	0,86	0,81	0,18	1,00	0,33	1,00	1,00	0,55
MDT	0,78	1,00	1,00	1,00	0,00	0,60	0,79	0,96	0,81	0,20	1,00	0,86	1,00	1,00	0,65
MRI	0,78	1,00	1,00	1,00	0,32	0,94	0,83	0,79	0,74	0,14	1,00	0,53	1,00	0,96	0,67
MNN	0,79	1,00	1,00	1,00	0,85	0,56	0,85	0,96	0,81	0,00	1,00	0,33	0,98	1,00	0,66
MSVM	0,77	0,99	1,00	1,00	0,04	0,88	0,85	0,96	0,81	0,21	1,00	0,48	0,98	1,00	0,62

Table 9 includes the precision of fault isolation based on the third scheme (Fig. 6). This scheme of fault isolation (Fig. 6) was divided into two parts. The first part was dealt with the selection of the basic classifiers, applied to isolate a single fault. After the analysis of the results presented in Table 7 the authors nominated the classifiers for single fault detection. These classifiers were chosen on the basis of general results. In case more than one classifier had the same efficiency value (more classifiers with the efficiency equal to 1, 000 for the fault at the same time) the authors pointed out a classifier with more stable results in the time domain. The accuracy and sensitivity of meta-classifiers are presented in Table 8. The accuracy of this method is a little bit higher than in the first scheme of fault isolation (Fig. 4) but the sensitivity (Table 8) and precision (Table 9) for the third scheme are more stable and similar for each class.

Table 9. The precision of fault isolation for meta-classifiers with a bank of classifiers for isolating single faults

	F1	F2	F7	F8	F10	F11	F12	F13	F14	F15	F16	F17	F18	F19
MkNN	1,00	1,00	1,00	0,35	0,47	0,95	0,72	0,96	0,36	1,00	0,57	1,00	0,84	0,91
MNB	1,00	1,00	1,00	0,32	0,44	0,69	0,70	0,99	0,30	1,00	0,84	1,00	0,84	0,75
MDT	0,99	1,00	1,00	0,00	0,82	0,70	0,72	0,95	0,40	1,00	0,34	1,00	0,84	0,90
MRI	1,00	1,00	1,00	0,38	0,55	0,70	0,94	0,96	0,36	1,00	0,41	1,00	0,81	0,89
MNN	1,00	1,00	0,75	0,31	0,84	0,96	0,72	1,00	0,00	1,00	0,99	1,00	0,84	0,86
MSVM	1,00	1,00	1,00	0,34	0,34	0,93	0,72	0,98	0,40	1,00	0,56	1,00	0,84	0,91

Five of the six single classifiers used in the first fault isolation process reached results between $0,67$ and $0,78$. One classifier is useless (decision tree) and its accuracy is equal to $0,07$. Meta-classifiers were able to significantly increase the accuracy of fault isolation. In this case the accuracy values of all classifiers are between $0,78$ and $0,87$. The sensitivity and precision values are also more stable and similar in each class. The last approach based on the third scheme of fault isolation (Fig. 6) shows also small improvement compared to the first scheme (see Fig. 4). Moreover, in this scheme the general efficiency of fault isolation is close to the result achieved by means of the single classifier. Generally, meta-classifiers in fault-isolation schemes are more stable and their results are similar to each other. In case of the second and third schema it is possible to use classifier based on decision tree so data prepared by base classifiers are simpler and more varied.

In this study, the authors used a confusion matrix in order to evaluate fault diagnosis systems that were created applying different classification schemes. Nevertheless, the accuracy, sensitivity and precision obtained from a confusion matrix can be directly compared with false and true detection/isolation rates proposed by the authors of the DAMADICS simulator [2]. The results of fault detection and isolation using single or meta-classification strategies that were achieved in this study are comparable to even more advanced methods described in the literature [8,34]. Furthermore, in this study the whole set of potential faults were investigated, whereas in the related papers only selected states were taken into consideration.

5   Conclusion

In the paper the application of selected classification schemes for fault diagnosis of the actuator systems was presented. The main purpose of the paper was to compare single and meta-classification strategies that could be successfully used as knowledge representation in the diagnostic expert system that is realized within the frame of the DISESOR project. The research was carried out basing on the well-practised hard and soft computing classification methods. The current paper can be viewed as the extension of the research work presented in [20]. In this study, the authors proposed a new approach for searching proper values of relevant parameters of classifiers used to fault diagnosis. The examined methods were tested in the context to be used for designing the engine of the DISESOR system. The comparison study was carried out within the DAMADICS benchmark problem. The classification schemes were implemented in RapidMiner software which is a well-known open source system for data mining and knowledge discovery. The particular results of the fault detection procedures showed that for simple industrial actuators it is possible to apply simple classification schemes without the necessity of using more advanced methods which are based on meta-classifiers. The final results reached in this paper are much better than results showed in [20]. The features and parameters of classifiers can be automatically tuned to increase their accuracy and sensitivity.

Overall, the application of single or meta-classification strategies with optimizing of relevant parameters allows to create effective as well as relatively less-complicated computational fault detection and isolation systems that can be successfully employed for on-line and off-line fault diagnosis of industrial actuators.

Acknowledgement. The research presented in the paper was partially financed by the National Centre of Research and Development (Poland) within the frame of the project titled "Zintegrowany, szkieletowy system wspomagania decyzji dla systemów monitorowania procesów, urządzeń i zagrożeń" (in Polish) carried out in the path B of Applied Research Programme - grant No. PBS2/B9/20/2013. The part of the research was also financed from the statutory funds of the Institute of Fundamentals of Machinery Design.

References

1. Akthar, F., Hahne, C.: RapidMiner 5, Operator Reference (2012). http://www. rapid-i.com
2. Bartyś, M., Patton, R., Syfert, M., de las Heras, S., Quevedo, J.: Introduction to the DAMADICS actuator FDI benchmark study. Control Eng. Pract. **14**(6), 577–596 (2006). http://dx.doi.org/10.1016/j.conengprac.2005.06.015
3. Bartyś, M., Syfert, M.: Using damadics actuator benchmark library (dablib). https://www.researchgate.net/publication/272478998_Using_damadics_actuator_benchmark_library_dablib
4. Bhadra, T., Bandyopadhyay, S., Maulik, U.: Differential evolution based optimization of svm parameters for meta classifier design. Procedia Tech. **4**, 50–57 (2012). http://dx.doi.org/10.1016/j.protcy.2012.05.006
5. Blanke, M., Kinnaert, M., Lunze, J., Staroświecki, M.: Diagnosis and Fault-Tolerant Control. Springer, Heidelberg (2006)
6. Breiman, L., Friedman, J.H., Olshen, R.A., Stone, C.J.: Classification and Regression Trees. Wadsworth, Belmont, CA (1984)
7. Caccavale, F., Villani, L.: Fault Diagnosis and Fault Tolerance for Mechatronic Systems: Recent Advances. Springer Tracts in Advanced Robotics. Springer, Heidelberg (2003)

8. Calado, J., Sá da Costa, J., Bartyś, M., Korbicz, J.: FDI approach to the damadics benchmark problem based on qualitative reasoning coupled with fuzzy neural networks. Control Eng. Practice **14**(6), 685–698 (2006). http://dx.doi.org/10.1016/j.conengprac.2005.03.025
9. Cholewa, W.: Real-time diagnostic expert systems. CAMES **9**(1), 21–40 (2002)
10. Cichosz, P.: Systemy uczące się. Warszawa: WNT (2000)
11. Cohen, W.W.: Fast effective rule induction. In: Twelfth International Conference on Machine Learning (1995)
12. Cretulescu, R., Morariu, D., Breazu, M., Vintan, L.: Weights space exploration using genetic algorithms for meta-classifier in text document classification. Stud. Inf. Control **21**(2), 147–154 (2012)
13. Rutkowska, D., Piliński, M., Rutkowski, L.: Sieci neuronowe, algorytmy genetyczne i systemy rozmyte. Wydawnictwo Naukowe PWN (1997)
14. Guyon, I., Elisseeff, A.: An introduction to variable and feature selection. J. Mach. Learn. Res. **3**, 1157–1182 (2003)
15. Haykin, S.: Neural Networks: A Comprehensive Foundation, 2nd edn. Prentice Hall International, Upper Saddle River (1999)
16. Hsu, C.W., Chang, C.C., Lin, C.J.: A practical guide to support vector classification. Technical report, Department of Computer Science, National Taiwan University (2003). http://www.csie.ntu.edu.tw/~cjlin/papers.html
17. Isermann, R.: Model-based fault detection and diagnosis - status and applications. Ann. Rev. Control **29**(1), 71–85 (2005). http://dx.doi.org/10.1016/j.arcontrol.2004.12.002
18. Isermann, R.: Fault-Diagnosis Systems: An Introduction from Fault Detection to Fault Tolerance. Springer, Heidelberg (2006)
19. Jingyuan, T., Yibing, S., Longfu, Z., Wei, Z.: Analog circuit fault diagnosis using adaboost and svm. In: Communications, Circuits and Systems, pp. 1184–1187 (2008). http://dx.doi.org/10.1109/ICCCAS.2008.4657978
20. Kalisch, M., Przystałka, P., Timofiejczuk, A.: Application of selected classification schemes for fault diagnosis of actuator systems. In: Proceedings of the 2014 Federated Conference on Computer Science and Information Systems, pp. 1381–1390 (2014). http://dx.org/10.15439/2014F158
21. Kerdprasop, K., Kerdprasop, N.: Feature selection and boosting techniques to improve fault detection accuracy in the semiconductor manufacturing process. In: Proceedings of the International MultiConference of Engineers and Computer Scientist (2011)
22. Korbicz, J., Kościelny, J.M., Kowalczuk, Z., Cholewa, W. (eds.): Fault Diagnosis. Models, Artificial Intelligence, Applications. Springer, Berlin (2004). http://dx.doi.org/10.1017/S0263574704241133
23. Korbicz, J., Kowal, M.: Neuro-fuzzy networks and their application to fault detection of dynamical systems. Eng. Appl. Artif. Intell. **20**(5), 609–617 (2007). http://dx.doi.org/10.1016/j.engappai.2006.11.009
24. Kościelny, J.M.: Diagnostyka zautomatyzowanych procesów przemysłowych. Akademicka Oficyna Wydawnicza EXIT, Warszawa (2001)
25. Kuncheva, L.: Combining Pattern Classifier: Methods and Algorithms. Wiley-Interscience, New Jersey (2004)
26. Lam, L.: Classifier combinations: implementations and theoretical issues. In: Kittler, J., Roli, F. (eds.) MCS 2000. LNCS, vol. 1857, p. 77. Springer, Heidelberg (2000). http://dx.doi.org/10.1007/3-540-45014-9_7

27. Lile, A.: Analyzing e-learning systems using educational data mining techniques. Mediterr. J. Soc. Sci. **2**(3), 403–419 (2011). http://dx.doi.org/10.5901/mjss.2011.v2n3p403
28. Moczulski, W.: Inductive acquisition of diagnostic knowledge for states tree with complex structure. Mech. Syst. Signal Process. **15**(4), 813–825 (2001). http://dx.doi.org/10.1006/mssp.2001.1389
29. Mrugalski, M., Witczak, M., Korbicz, J.: Confidence estimation of the multi-layer perceptron and its application in fault detection systems. Eng. Appl. Artif. Intell. **21**, 895–906 (2008). http://dx.doi.org/10.1016/j.engappai.2007.09.008
30. Namdari, M., Jazayeri-Rad, H., Hashemi, S.J.: Process fault diagnosis using support vector machines with a genetic algorithm based parameter tuning. J. Autom. Control **2**(1), 1–7 (2014)
31. Patton, R., Uppal, F., Lopez-Toribio, C.: Soft computing approaches to fault diagnosis for dynamic systems: a survey. In: IFAC Symposium SAFEPROCESS, pp. 298–311, June 2000
32. Patton, R.J., Frank, P.M., Clark, R.N.: Issues of Fault Diagnosis for Dynamic Systems. Springer, Heidelberg (2000)
33. Pöyhönen, S.: Support vector machines in fault diagnostics of electrical motors. Helsinki University of Technology Control Engineering Laboratory, Technical report (2002)
34. Previdi, F., Parisini, T.: Model-free actuator fault detection using a spectral estimation approach: the case of the damadics benchmark problem. Control Eng. Pract. **14**, 635–644 (2006). http://dx.doi.org/10.1016/j.conengprac.2005.04.001
35. Puig, V., Witczak, M., Nejjari, F., Quevedo, J., Korbicz, J.: A GMDH neural network-based approach to passive robust fault detection using a constraint satisfaction backward test. Eng. Appl. Artif. Intell. **20**, 886–897 (2007). http://dx.doi.org/10.1016/j.engappai.2006.12.005
36. Moczulski, W.: Diagnostyka techniczna. Metody pozyskiwania wiedzy. Wydawnictwo Politechniki Śląskiej, Gliwice (2002)
37. Woźniak, M.: Metody fuzji informacji dla komputerowych systemów rozpoznawania. Oficyna Wydawnicza Politechniki Wrocawskiej (2006)
38. Wu, Q., Ni, Z.: Car assembly line fault diagnosis based on triangular fuzzy support vector classifier machine and particle swarm optimization. Expert Sys. Appl. **38**, 4727–4733 (2011). http://dx.doi.org/10.1016/j.eswa.2010.08.099
39. Yang, B.S., Di, X., Han, T.: Random forests classifier for machine fault diagnosis. J. Mech. Sci. Tech. **22**, 1716–1725 (2008). http://dx.doi.org/10.1007/s12206-008-0603-6
40. Yao, P., Liu, Z., Wang, Z., Bu, S.: Fault signal classification using adaptive boosting algorithm. Elektronika ir Elektrotechnika **18**(8), 97–100 (2012). http://dx.doi.org/10.5755/j01.eee.18.8.2635

Intelligent Association Rules for Innovative SME Collaboration

Gulgun Kayakutlu[1](✉), Irem Duzdar[2], Eunika Mercier-Laurent[3],
and Bahar Sennaroglu[4]

[1] Industrial Engineering Department, Istanbul Technical University,
34367 Macka, Istanbul, Turkey
kayakutlu@itu.edu.tr
[2] Industrial Engineering Department, Istanbul Arel University,
34537 Tepekent, Istanbul, Turkey
iremduzdar@arel.edu.tr
[3] MODEME, Centre Magellan, Jean Moulin University Lyon 3, Lyon, France
e.mercier-laurent@univ-lyon3.fr
[4] Marmara University, Istanbul, Turkey
bsennaroglu@gmail.com

Abstract. SMEs are encouraged to collaborate for research and innovation in order to survive in tough global competition. Even the technology SMEs with high knowledge capital have the fear to collaborate with other SMEs or bigger companies. This study aims to illuminate the preferences in customer, supplier and competitor collaboration within industry or inter industry. A survey is run on more than 110 companies and Machine Learning methods are used to define the association rules that will lead for success.

Keywords: Collaborative innovation · Association rules · SVM · SOM

1 Introduction

Knowledge based SMEs need to construct successful alliances in order to have sustainable business in a competitive environment. Global experiences with randomly chosen collaborators have shown failures that caused the fear of new collaborative work. Causes of failure based on the culture and the type of collaboration are studied [1]. The collaboration and creativity must be in the required skills to develop an effective knowledge management [2, 3]. Abereijo et al. states that the innovation capability of any company depends on its capacity to combine the information assets and the experiences which can be found in the literature about the national innovation studies [4]. The approaches of owners to innovation, their tendencies and enthusiasm, and the social activities in the company, and dependence among workers are very important factors for effective knowledge management in SMEs [5]. Alliance in new product development has been the focus of industrial researchers [6–8].

One of the effective tools in knowledge management is becoming the association rules. There are many practices in technology firms and development projects for new products, but it is not used widely for cooperation of companies yet. Basically, to

© IFIP International Federation for Information Processing 2015
E. Mercier-Laurent et al. (Eds.): AI4KM 2014, IFIP AICT 469, pp. 150–164, 2015.
DOI: 10.1007/978-3-319-28868-0_9

allocate the resources efficiently, the rules must be stated to regulate the information transfer. The well designed rules will be very effective to get benefits to SMEs which have limited capital and sources to continue the innovative operations. It can be seen in literature that the achievement measures and performance studies are useful for innovative collaboration. In business world, these criteria is trying to state a strategy for an individual company without employing the information coming from the knowledge assets of worldwide applications by decision makers.

This study aims to provide a pre-analysis of the path for successful alliances that will lead improvements in innovative power. Both qualitative and quantitative analysis of the SME alliances is realized to find the conditions causing failures and supporting the success in innovation. Support Vector Machine and Self Organized Maps are used to define the most frequent patterns that will give the support and confidence to identify the relationships. Association rules achieved will determine the optimal use of resources.

This paper is so organized that the literature review will be given in the second section and the methodology definition will follow. The fourth section will be reserved for presenting the survey and the results. The conclusion will be given in the fifth and last section.

The implication of the study is generic enough to help any SME or research organization or large business to reduce risks in future alliances.

2 Background

The surveys on the interactions of the industrial stockholders had begun at the early 21st century. It can be seen easily that the expansive applications of outsourcing increases the importance of this matter.

More than 70 publications have been examined to have an understanding about the applications of the rules for cooperation between SMEs aimed at innovation. These studies implied that the main attention was on the various natures of interactions with SMEs and with Large Companies, either in an identified industry or different industries; either only for customers or both of the customers and suppliers. At the end of the detailed examination of this topic, the number of analyzed researches was 112, which were useful to classify on the base of methods.

The first research on Association Rule Mining and Methods is found in 1996 [9] trying to find the most frequent occurrences of events to support the linked processes. The research in the field followed the timeline shown in Fig. 1.

The analysis in engineering branch shows that beginning from 2005, it takes its place increasingly reached 8 researches on annual average. The 25 % of these researches were done in 2009, and the maximum number is reached in this year.

The researches on association rule started in 1997 and the number of publications increased until 2007, most of the studies were done about general subjects in this period. Then the general purpose studies reduced because the specification starts (Fig. 2). The annual distribution of the published studies on 'Association Rule Methodology in Data Mining' is researched. The number of published applications about this subject increased until 2006. Between 2009–2011 years there is slight decrease, and then, last 3 years an increasing trend is observed.

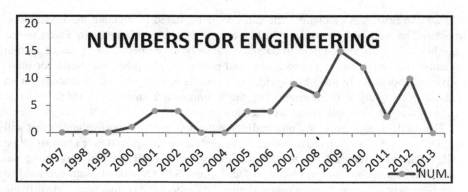

Fig. 1. Research timeline on association rules

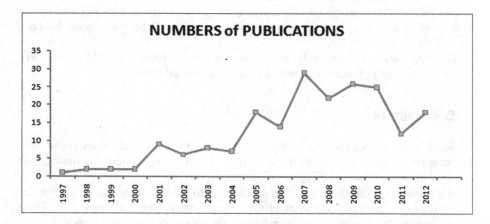

Fig. 2. Publication timeline on association rules

The usage of association rule in engineering is analyzed and the numbers of publications, the starting time seems 1999, but the reel beginning time is 2005. This matches the starting time of specialization on association rule studies.

Kakabadse et al. showed that using improved communication and information technologies will improve the SME collaboration [1]. Also Post et al. showed the fact that SMEs would like to collaborate only for developing new products [10]. New product based collaboration has evolved fast [8]. Corporation and competition are found as flaming collaboration types that feed the SME improvements in innovation [7]. Association rules defined for the failure types have opened a new dimension for the research on failure of collaboration [11]. The first study on mining the SME innovation by Wang et al has found some patterns for allocating the R&D resources [12]. Suh and Kim have detailed the R&D collaboration in service industries detected the positive relations of technology and the product or process innovation [13]. Swarnkar et al. analyzed when and how the collaboration strategies will be used in virtual organizations

[14]. Wiltsey et al. claimed that extent, nature or impact of R&D programs are studied rarely. The interactions among the influences must be given in multiple levels and fidelity and changes must be observed in time [15]. Woodland and Hutton introduced the social dimension on the collaborative success [16]. Both the fear issues and the success causes studied by Bouncken et al. defined technology influencers, sharing the knowledge and learning from the partner as the main influencers [6]. Bayo-Moriones et al. (2013) proposed that the effects of information and communication technology (ICT) has an indirect on the achievement of increasing the marketing capacity, profitability, and profit rates because of the advanced internal and external communication channels; just like the developed operational processes [17]. Sawers et al. concluded that the intrinsic advancement level of SMEs has inverse correlation with the enthusiasm for collaboration, where the effect of extrinsic level has direct correlation [18]. Rocha showed that empowering the business and the high level of competition at market area will lead to a larger SME industry [19]. Franco and Haase introduced various collaborating businesses and stated their general policies, and tendencies to cooperation together with their courses of action [20]. Antlová et al. insisted that the companies and their knowledge assets depending on the basic qualifications exhibit the developing businesses [21].

Knowledge management and data mining overviews [22] and Knowledge Management performance studies [23] realized recently do not show any association rule study for the collaborative innovation success and failure.

3   Methodologies

3.1   Association Rules

Given a set of transactions, rules are defined that will exhibit that the occurrence of an item based on the occurrences of other items in the transaction. This is the *association analysis*. It is useful to explore the interesting relations, which are embedded in the huge data sets. These hidden interactions can be stated in the form of *association rules* [24]. The strength of an association rule is measured with its support and confidence values. Support shows the how often that rule is applicable to a given dataset. The Confidence is the occurrence frequency of the item in that transaction [9].

Support (s) is the fraction of transactions that contain an itemset

$$s(X \rightarrow Y) = \frac{\sigma(X \cup Y)}{N} \tag{1}$$

Confidence (c) measures how often items in Y appear in transactions that contain X

$$c(X \rightarrow Y) = \frac{\sigma(X \cup Y)}{\sigma(X)} \tag{2}$$

The item set patterns are found in various methods which could be apriori or aposteriori. The overview of all the methods used in association srule studies are given in Fig. 3.

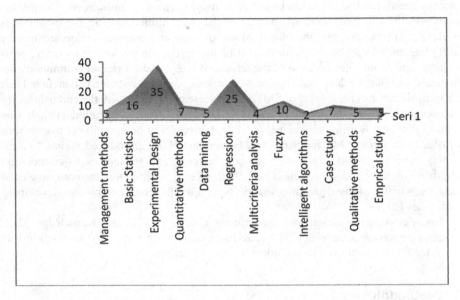

Fig. 3. Overview of association rule techniques

3.2 Support Vector Machines

It is a machine learning technique, which is mainly introduced for classification in two classes [25] but further used in clustering [26].

It can be analyzed as an optimization problem as in Eq. 3 [27] relaxed with Lagrange multipliers in objective function as in Eq. 4.

$$
\begin{aligned}
&\min z = \tfrac{1}{2}\|w\|^2 \\
&s.t. \\
&r_i(w_i x_i + w_0) \geq 1
\end{aligned}
\tag{3}
$$

Data is separated with a hyper-plane multiplied by -1 or $+1$.

$$
L_p = \frac{1}{2}\| w \|^2 - \sum_i \lambda_i r_i (w_i x_i + w_0) + \sum_i \lambda_i
\tag{4}
$$

In 1990s Vapnik pioneered the development of support vector machines specifically for classification in two classes [25]. This method aims to develop an optimal hyper plane as a decision function using the maximum distance among two class vectors. Support Vector Using a Gaussian Kernel as defined in Eq. 5 will increase the reliability on dissimilarities [29].

$$K(x, x_i) = \exp\left(-\frac{\|x - x_i\|^2}{2}\right) \tag{5}$$

3.3 Self Organized Maps

Self-Organizing Map (SOM) is a widely used artificial neural network technique in clustering with unsupervised learning algorithm. This technique clusters according to the similarities to the input data [30]. SOMs structure the output with individual node similarity as well as cluster center distance. This technique is based on competitive learning, where the output nodes are made of the winning node activated by one input node. The output nodes would have scoring values using a function, most commonly Euclidean distance between the inputs and weights. For each input vector x, and for each output node j, the value D (w_j, x_n) of the scoring function. Euclidean distance function is shown in Eq. 6.

$$Dw_j, x_n = \left(w_{ij} - x_{ni}\right)^2 \tag{6}$$

The winning node therefore becomes the center of a neighborhood of excited nodes. In self- organizing maps, all nodes in the given neighborhood share competition. Therefore, even if the nodes in the output layer are not connected directly to the input layer, they tend to share common features, of the neighborhood [31]. The nodes in the neighborhood of the winning node participate in adaptation, which is, learning. The weights of these nodes are adjusted to improve the weights defined in Eq. 7, until a threshold is reached.

$$w_{ij} \; new = w_{ij} \; current + \alpha x_{ni} - w_{ij} \; current \tag{7}$$

In Eq. 7 α is the learning rate. If it is necessary, the learning rate and neighborhood size are adjusted.

4 Application

A survey is run with the technology firms sited in Techno-parks of linked, 5 are about competences and 4 the technology choices. 130 firms responded but only 105 are included in the analysis. 14 % of the companies were medium size and 37 % of them were aged more than 10 years. They have chosen the type of collaboration among the SME and Big firms as well as among the customers, suppliers and competitors as shown in Fig. 4. The reason for innovative collaboration is stated as shown in Fig. 5.

When the choices of SMEs for innovation based collaboration is examined, the results are as follows. The number of responses is greater than 105, since this question

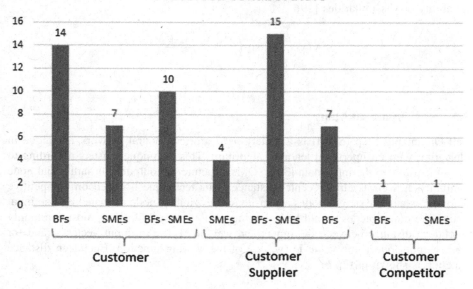

Fig. 4. Choice of collaboration according to size and relation

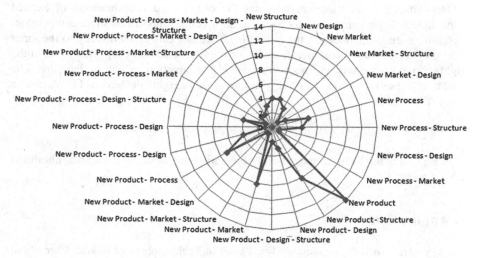

Fig. 5. Innovation causes in collaboration

gives the choice to select more than 1 reason for innovation based collaboration. Most of the SMEs are trying to develop a new product through innovation investments; total number is 61. Secondly 35 firms are trying to have a new process at the end of the innovation tasks. There are 33 firms trying to find new design. The number of firms trying to reach new markets is 25. The number of SMEs targeting the structural change is 23. 19 of the firms have specified different goals. Just three of the firms aim all of the goals that are mentioned here.

SVM is applied to classify the collaboration made specifically for innovation, validated by Cronbach alpha, that resulted as more than 0.7. SOM is used to cluster those to show the supporting frequencies of choices. Then the cross tables are achieved as a basis for the association rules. The first achievement was micro companies would like all the collaborators to have innovation culture which is less important for the bigger companies. Some other samples of cross tables are given as below.

Table 1 gives the fact that majority of responders prefer for at least one collaborator to have design competence.

Table 1. Business understanding and design competence

Innov			Design		
			All	At least one	Neglige
Collaboration for others	Bus-under	All	4	14	
		At least one	2	15	
Collaboration for innovation	Bus-under	All	14	26	4
		At least one	7	13	2
		Neglig	1	1	2

In the firms collaborating for innovation for 1 to 3 years for the success of innovation it is necessary understanding the market requirements by all the firms together with the well-developed innovation culture (Table 2).

Table 2. Age-innovation culture and market requirements relation

COLLAB_Age			INNO_CULT		
			All	At least one	Neglige
0	UNDERSTD_REQ	All	13	8	
		More than one	6	6	
		At least one	0	1	
<1 year	UNDERSTD_REQ	All	9	1	0
		More than one	3	1	1
1–3 years	UNDERSTD_REQ	All	13	1	
		More than one	1	1	
		At least one	4	2	
3–5 years	UNDERSTD_REQ	All	6	1	
		More than one	3	2	
		At least one	0	1	
>5	UNDERSTD_REQ	All	8	2	0
		More than one	3	2	2
		At least one	2	2	0

5 Results

The Reliability analyzes have been done on the results of the questionnaires, the Cornbach's Alpha value is 0.602; this value is in the acceptable range. In the logistic regression model, the determined significance level is 0.100. The values of attributes of innovation are neglected, since they are greater than 0.100. Innovation related criteria in this study have no significance based on firm size and collaborator types of these firms. In other words, firms care the technological features but ignore the innovative attributes. In the table below, the significance values are shown for the other 3 attributes and the results of analyzes for features of the questions. The significance values less than 0.100 are used here (Fig. 6).

Significance Table				
		Clusters		
		Finance	Technology	Management
Properties of Firms				
	Firm Size	-	-	-
	Employee Size	0.000	0.000	0.000
Firm Size	Firm Age	0.007	0.003	0.045
	Collaborate for Innovation	0.111	0.068	0.399
	Collaboration Duration	0.019	0.019	0.027
	Firm Size	0.000	0.000	0.000
	Employee Size	-	-	-
Employee Size	Firm Age	0.152	0.193	0.074
	Collaborate for Innovation	0.497	0.441	0.083
	Collaboration Duration	0.560	0.513	0.013
	Firm Size	0.019	0.000	0.000
	Employee Size	0.176	0.002	0.009
Firm Age	Firm Age	-	-	-
	Collaborate for Innovation	0.203	0.002	0.030
	Collaboration Duration	0.000	0.000	0.000
	Firm Size	0.006	0.019	0.041
	Employee Size	0.054	0.076	0.084
Collaborate for Innovation	Firm Age	0.010	0.025	1.000
	Collaborate for Innovation	-	-	-
	Collaboration Duration	0.000	0.000	0.000
	Firm Size	0.099	0.001	0.000
	Employee Size	0.624	0.045	1.000
Collaboration Duration	Firm Age	0.000	0.000	0.043
	Collaborate for Innovation	0.000	0.000	0.000
	Collaboration Duration	-	-	-

Fig. 6. Significance values for each clusters based on demographic properties

After the multinomial logistic regression analysis coefficients for all attributes are obtained. The statistically significant attributes are used. Coefficients of financially related criteria are shown in Fig. 7.

The cross relation tables have been constructed to define the rules obtained from the model.

Finance Related Criteria:
RULE 1:
IF
(Firm Size = "Micro" AND Firm Age ≤ "1")
THEN
(Innovation operation expenditure = "proportionally shared" AND Price = "important")

SIZE= 2		B	Std. Error	Wald	df	Sig.	Exp(B)	90% Confidence Interval for Exp(B)	
								Lower Bound	Upper Bound
	Intercept	-4.398	2.530	6.387	1	.011			
	[ComAge_A=0]	2.078	1.020	4.156	1	.041	7.994	1.494	42.771
	[ComAge_A=1]	0			0				
	[ComAge_B=0]	1.725	.758	5.186	1	.023	5.614	1.615	19.522
	[ComAge_B=1]	0			0				
	[ComAge_C=0]	.652	.680	.920	1	.338	1.920	.627	5.877
	[ComAge_C=1]	0			0				
	[ComAge_D=0]	-.782	.732	1.056	1	.304	.472	.142	1.571
	[ComAge_D=1]	0			0				
	[Cap_A=0]	.055	.611	.008	1	.928	1.057	.387	2.889
	[Cap_A=1]	0			0				
	[Cap_B=0]	.145	.718	.041	1	.840	1.156	.355	3.768
	[Cap_B=1]	0			0				
	[Innov_Op_A=0]	.184	.716	.066	1	.797	1.203	.370	3.906
	[Innov_Op_A=1]	0			0				
	[Innov_Op_B=0]	.741	.714	1.075	1	.300	2.098	.648	6.794
	[Innov_Op_B=1]	0			0				
	[Price_A=0]	-.504	.927	.296	1	.587	.604	.132	2.775
	[Price_A=1]	0			0				
	[Price_B=0]	.473	.973	.237	1	.627	1.606	.324	7.957
	[Price_B=1]	0			0				
	[Export_A=0]	2.347	.910	6.659	1	.010	10.459	2.342	46.706
	[Export_A=1]	0			0				
	[Export_B=0]	1.839	.792	5.389	1	.020	6.289	1.709	23.142
	[Export_B=1]	0			0				

Fig. 7. Regression coefficients & significancies

MEANING:

✓ *Preferences for the initiating micro SMEs emphasize the innovation operation expenditures according to the collaborator sharings and market value (price) of the innovated product (Figure 8).*

RULE 2:
IF
(Firm Size = "Small" AND Collaborate = "Large Firms")
THEN
(Capital = "more than average" AND Exportation facilities = "only one firm")

MEANING:

✓ *Small SMEs emphasize that capital of the collaborators are to be more than the sector average and exportation facilities are done by only one collaborator for innovation (Figure 9).*

RULE 3:
IF
(Firm Age = ">10 years old" AND Collaboration Duration = "3–5 years")
THEN
(Capital = "more than average" AND Price = "motivation")

Finance Related Criteria		Capital_More than Average of Sector (A)	Capital_Average of Sector (B)	Innovation Operation Expenditure_Balance (A)	Innovation Operation Expenditure_More is more (B)	Price_Important (A)	Price_Only motivation (B)	Exportation_All collaborators (A)	Exportation_Only one (B)
Firm Size	Firm Age								
MICRO (1)	<1 YEAR OLD (A)				■	■			
	1-3 YEARS OLD (B)				■	■			
	3-5 YEARS OLD ©								
	5-10 YEARS OLD (D)								
SMALL (2)	<1 YEAR OLD (A)								
	1-3 YEARS OLD (B)		■			■			
	3-5 YEARS OLD ©								
	5-10 YEARS OLD (D)		■				■		
MEDIUM (3)	<1 YEAR OLD (A)								
	1-3 YEARS OLD (B)								
	3-5 YEARS OLD ©							■	
	5-10 YEARS OLD (D)								

Fig. 8. Cross – relations Table: Firm Size – Firm Age – Finance Related Criteria

Finance related Criteria		Capital_More than Average of Sector	Capital_Average of Sector	Innovation Operation Expenditure_Balance	Innovation Operation Expenditure_More is more	Price_Important	Price_Only motivation	Exportation_All collab	Exportation_Only one
COLL	Firm Size								
Cust–LFs	Micro			■			■	■	
	Small					■			
	Medium								
Cust–SMEs	Micro								
	Small	■							
	Medium								■
Cust-Supp–LFs	Micro						■		
	Small								
	Medium								
Cust-Supp–SMEs-LFs	Micro						■		
	Small					■			■
	Medium								
Cust	Micro								
	Small					■			
	Medium								
LFs	Micro	■			■			■	
	Small								■
	Medium								
Cust-Supp-SMEs	Micro		■					■	
	Small								
	Medium								

Fig. 9. Cross – relations Table: Collaboration – Firm Size – Finance Related Criteria

MEANING:

✓ *10 years (and more) old SMEs emphasize capital of the collaborators have to be more than the sector average whereas market value (price) of the innovated product is only a motivation.*

Technology Related Criteria:
IF
(Firm Size = "Small" AND Firm Age = "3–5 years old")
THEN
(Connectivity = "All possible ways" AND Change Management = "All Collaborators")

MEANING:

✓ *SMEs with age 5 to 10 years old emphasize change management is to be applied by all collaborators and use all possible connectivity possibilities.*

Figure 10 shows the cross – relations between collaborator type (customer – supplier – SMEs – large firms) and firms' size for the technology related criteria. Also this mentioned model makes sense statistically significant as a result of the logistic regression analysis.

Technology Related Criteria		MIS_ Least one	MIS_All	Comm_If tech exist	All tech	ChngMng _individual	ChngMng_ Together	Connect _only one type	Connect _All ways
COLL	Firm Size								
	Micro								
Cust-LFs	Small								
	Medium								
	Micro								
Cust–SMEs	Small								
	Medium								

Fig. 10. Cross – relations Table: Collaboration – Firm Size – Technological Criteria

RULE 2:
IF
(Firm Size = "Small" years old" AND Collaborate = "Customers" and "LFs")
THEN
(Communication technologies = "All opportunities" AND Change Management = "Individual")

MEANING:

✓ *Small SMEs who are collaborating with the customers which are large firms (LFs) emphasize usage of all communication technologies opportunities and change management is can be individual choice (Figure 9).*

Management Related Criteria:
RULE 1:
IF
(Firm Size = "Medium" AND Firm Age = "5–10 years old")
THEN
(Professionalism = "Motivation" AND Organizational Structure = "Effective" AND Cooperation & Coordination = "All" AND Leadership = "Only one")

MEANING:

✓ *Medium size SMEs of age 5-10 years prefer to collaborate with companies which have effective organizational structure with both cooperation and coordination attitude; in the collaboration a single leader is preferred and professionalism can be taken only as the motivator.*

RULE 2:
IF
(Firm Size = "Micro" AND Collaborate = "Customers" and "SMEs")
THEN
(Professionalism = "All" AND Business Experience = "All" AND Leadership = "All")

MEANING:

✓ *The micro SMEs collaborating with the customers emphasize professionalism, business experience for the problem solving and they prefer all the collaborators to have the leadership features.*

6   Conclusion

This study investigates the most preferred conditions for a successful collaboration for innovative SMEs. SVM and SOM are used to construct the basis for creating the association rules. As the result of a survey in Turkey, there are hundreds of relations depicted in the analysis.

The achievements are interesting enough to show that the technology companies are confused in differentiating the technology and innovation concepts. It was interesting to observe micro and young companies not willing to collaborate with the big and overwhelming companies. Everybody asks for full communication technology, but only small SME with 5–10 years of experience ask for the collaborators to have effective organization and full professionalism.

The validation by logistic regression on the same data is in process. All the results achieved using logistic regression will be cross-validated with machine learning application results. Future survey will be aiming to improve the innovation concept of the technology firms in detail.

References

1. Kakabadse, N.K., Kakabadse, A., Ahmed, P.K., Kouzmin, A.: The ASP phenomenon: an example of solution innovation that liberates organization from technology or captures it? Eur. J. Innov. Manag. **7**(2), 113–127 (2004)
2. Horak, B.J.: Dealing with human factors and managing change in knowledge management: a phased approach. Top. Health Inf. Manag. **21**(3), 8–17 (2001)
3. Yahya, S., Goh, W.K.: Managing human resources toward achieving knowledge management. J. Knowl. Manag. **6**(5), 457–468 (2002)
4. Abereijo, I.O., Adegbite, S.A., Ilori, M.O., Adeniyi, A.A., Aderemi, H.A.: Technological innovation sources and institutional supports for manufacturing small and medium enterprises in Nigeria. J. Technol. Manag. Innov. **4**(2), 82–89 (2009)
5. Wee, J.C.N., Chua, A.Y.K.: The peculiarities of knowledge management processes in SMEs: the case of Singapore. J. Knowl. Manag. **17**(6), 958–972 (2013)
6. Bouncken, R.B., Kraus, S.: Innovation in knowledge-intensive industries: the double-edged sword of coopetition. J. Bus. Res. **66**(10), 2060–2070 (2013)
7. Gnyawali, D.R., Park, B.R.: Co-opetition and technological innovation in small and medium sized enterprizes a multilevel conceptual model. J. Small Bus. Manag. **47**(3), 308–330 (2009)
8. Narula, R.: R&D collaboration by SMEs: new opportunities and limitations in the face of globalisation. Technovation **24**(2), 153–161 (2004)
9. Agrawal, R.: Parallel mining of association rules. IEEE Trans. Knowl. Data Eng. **8**(6), 962–969 (1996)
10. Post, G.J.J., Hop, L., van Aken, J.E.: Indicators for establishing SME product development networks. J. Sci. Ind. Res. **60**(3), 264–276 (2001)
11. Zheng, Z., Lan, Z., Park, B.H., Geist, A.: System log pre-processing to improve failure prediction. In: 2009 IEEE/IFIP International Conference on Dependable System Networks (2009)
12. Wang, C.H., Chin, Y.C., Tzeng, G.H.: Mining the R&D innovation performance processes for high-tech firms based on rough set theory. Technovation **30**, 447–458 (2010)
13. Suh, Y., Kim, M.-S.: Effects of SME collaboration on R&D in the service sector in open innovation. Innov. Manag. Policy Pract. **14**(3), 349–362 (2012)
14. Swarnkar, R., Choudhary, A.K., Harding, J.A., Das, B.P., Young, R.I.: A framework for collaboration moderator services to support knowledge based collaboration. J. Intell. Manufact. **23**, 2003–2023 (2012)
15. Wiltsey Stirman, S., Kimberly, J., Cook, N., Calloway, A., Castro, F., Charns, M.: The sustainability of new programs and innovations: a review of the empirical literature and recommendations for future research. Implement. Sci. **7**, 17 (2012)

16. Woodland, R.H., Hutton, M.S.: Evaluating organizational collaborations: suggested entry points and strategies. Am. J. Eval. **33**, 366–383 (2012)
17. Bayo-Moriones, A., Billón, M., Lera-López, F.: Perceived performance effects of ICT in manufacturing SMEs. Ind. Manag. Data Syst. **113**(1), 117–135 (2013)
18. Sawers, J.L., Pretorius, M.W., Oerlemans, L.A.G.: Safeguarding SMEs dynamic capabilities in technology innovative SME-large company partnerships in South Africa. Technovation **28**(4), 171–182 (2008)
19. Rocha, E.A.G.: The impact of the business environment on the size of the micro, small and medium enterprise sector; preliminary findings from a cross-country comparison. Procedia Econ. Finance **4**, 335–349 (2012)
20. Franco, M., Haase, H.: Interfirm alliances: a taxonomy for SMEs. Long Range Plann. (2013)
21. Antlová, K., Popelínsky, L., Tandler, J.: Long term growth of SME from the view of ICT competencies and web presentations. E + M Ekonomie a management **1**, 125–138 (2011)
22. Tsai, H.H.: Knowledge management vs. data mining: research trend, forecast and citation approach. Expert Syst. Appl. **40**(8), 3160–3173 (2013)
23. Kim, T.H., Lee, J.N., Chun, J.U., Benbasat, I.: Understanding the effect of knowledge management strategies on knowledge management performance: a contingency perspective. Inf. Manag. **51**, 398–416 (2014)
24. Jackson, J.: Data mining: a conceptual overview. Commun. Assoc. Inf. Syst. **8**, 267–296 (2002)
25. Cortes, C., Vapnik, V.: Support vector networks. Mach. Learn. **20**, 273–297 (1995)
26. Finley, T., Joachims, T.: Supervised clustering with support vector machines. In: Proceedings of the 22nd International Conference on Machine Learning (ICML), pp. 217–224 (2005)
27. Haykin, S.: Neural Networks: a Comprehensive Foundation. Prentice-Hall, New Jersey (1999)
28. Alpaydın, E.: Machine Learning. Massachusetts Institute of Technology, USA (2004)
29. Cherkassky, V., Ma, Y.: Practical selection of SVM parameters and noise estimation for SVM regression. Neural Netw. **17**, 113–126 (2004)
30. Leopold, E., May, M., Paaß, G.: Data mining and text mining for science & technology research. In: Moed, H.F., Glänzel, W., Schmoch, U. (eds.) Handbook of Quantitative Science and Technology Research - The Use of Publication and Patent Statistics in Studies of S&T Systems, pp. 187–213. Springer, The Netherland (2004)
31. Larose, D.T.: Discovering Knowledge in Data: An Introduction to Data Mining. John Wiley & Sons Inc, New Jersey (2005)

Managing Intellectual Capital in Knowledge Economy

Eunika Mercier-Laurent[(⊠)]

University Jean Moulin Lyon 3, Cours Albert Thomas, 69008 Lyon, France
eunika.mercier-laurent@univ-lyon3.fr

Abstract. Strategic Knowledge Management considers Intellectual Capital (IC) as roots of all organizations activities. The success of organizations strongly depends on the way they manage all facets of knowledge and skills. Artificial Intelligence brought some methods and techniques for handling intellectual assets of companies, expertise management, knowledge transfer and training. This paper presents an overview of experiences and research in applying artificial intelligence approaches and techniques for intellectual capital management and gives some perspectives for the future.

1 Introduction

Since over two decades the interest for managing intangible assets, including intellectual capital has been grown. However the roots of intellectual capital go far back in the history. Skyrme [24] mentions seven ages of Knowledge Management beginning in 1970s. The know-how has been shared by doing or by storytelling. In XX century the term of human capital has been probably re-introduced by the economist Theodore Schultz in 1961 [1]. He considers that the investment in human capital is crucial for the economic development and the education has a key contribution. Latter the term of "intellectual capital" have been introduces to cover larger field including patents and documents. Among training professionals, Sveiby [2] defined an Intangible Assets Monitor to drive the management of human capital. Many reports on Intellectual Capital have been published and numerous databases have been implemented [26]. Reports provide static information that should be updated, it is the same for databases; the both are not adapted for smart managing of human resources, especially in the globalization context.

Probably the first effort in applying the artificial intelligence techniques to managing skills in given situation was the application developed for French police [3]. These principles and architecture were reused in larger system for managing security of the Winter Olympic Games [4]. The techniques such as case-based reasoning can also be useful for matching demand and offer (looking for a job or a skill).

Organizations such as OECD [5] have been involved in defining a general methodology for measuring intangible investment since 1989.

The globalization changed the game of economic development. Intellectual capital is said to become an important asset and its assessment and management has turned to a priority for the Knowledge Economy. The intellectual capital is among the hot topics of

© IFIP International Federation for Information Processing 2015
E. Mercier-Laurent et al. (Eds.): AI4KM 2014, IFIP AICT 469, pp. 165–179, 2015.
DOI: 10.1007/978-3-319-28868-0_10

conversations, conferences, magazines, scientific journals, books and reports. However companies and organizations are still measuring their success in term of financial capital and ROI (return on investment). Corporate Social Responsibility (CSR) recommends using local resources; it could trigger a better management of local talents.

This paper presents key references related to the evolution of human and intellectual capital, gives some elements of economic and environmental context and mentions current efforts in evaluating, measuring and managing of intellectual capital. It is followed by a presentation of a method and tools to manage this wealth differently and to stimulate a reflection on the role of this capital in the Knowledge Economy and in the Innovation Ecosystems.

2   State of the Art

The issue of intellectual capital is complex. It includes human capital and related relational capital as well as "paper and digital" capital (patents, books, papers, images, drawings…). The management of human capital involves various fields such as management, psychology, economy, sociology, communication, health, wellbeing, intellectual property rights and recently sustainable development. Intellectual capital forms the basis of the successful and sustainable development of companies, cities, regions and countries. Such a development requires the right way of managing the intangible wealth in connection with tangible ones and continuous search and exploring of opportunities.

Numerous publications provide a multidisciplinary view of the subject. According to Schultz [1], in charge of economic development, the education is the most important in managing of "human capital". Another economist Becker [6] considers education, training, and health as the major investments in human capital.

According to Dixon [9], training, capacity building and learning are key enabling factors for "sustainability" seen as long term ability of individuals and organizations to produce innovations as a reaction and adaptation to changes in external conditions. It is the link between opportunities, projects, addressing the real needs, and building capacity or empowerment that ensures useful learning, innovation and an economically efficient process. Training supports the development of all phases of the project lifecycle (situation analysis, forecasting, planning, implementation and evaluation/measurement of impacts). Trained persons develop skills and produce methods, information and knowledge required for the success of the project. Training, combined with the development and implementation of projects on the local level, allows: (i) increasing and mobilizing human and social capital (ii) developing new activities and (iii) creating interactions leading to collective dynamics to the invention of new rules and standards (institutional capital) needed to integrate new activities in formal economy.

Edvinsson and Malone [7] points out the role of intellectual capital in the modern economy and suggests adding to annual reports of firms a part on intangible values. Stahle et al. [27] state that having a world database of Intellectual Capital may improve the way it is managed and influence the welfare of related countries.

OECD [8] highlights the role of human capital in the development and well-being of nations.

Berkes et al. [10] propose to develop an "adaptive capacity". The concept has been used in biology and in the context of climate change, but applies to a much broader range of issues. Adaptive capacity developed in poor countries is very strong; it is extremely important capacity to be successful in XXI century. Persons able to adapt and to solve problems using individual and collective knowledge, as well as solutions from the past that work for current challenge, is able to survive and even lead in global dynamics – Mercier-Laurent [11]. Viability theory of Aubin [12] may be useful to control the balance of the ecosystems based on human capital as engine.

According to Savage [13] the 5th generation of managerial methods has to consider knowledge as asset. This statement has been enhanced by Amidon [14] in The Innovation Strategy for the Knowledge Economy. *To know* is the opposite of *to have* attitude cultivated in today world and focus exclusively on quick business. From education point of view the most important is to learn how and what to learn. These few references cover a large spectrum on human and intellectual capital themselves and the roles they play in economic development and the wellbeing of the nations.

From management point of vie, the connection between Balanced Scorecard and Intellectual Capital is studied [30].

Considering Information Technology (IT) and Information and Communication Technology (ICT), companies have now human resource databases, but in many cases they contain just basis information on employees and are not often updated. There is some tries to use the Enterprise Social Networks (ESN) as a source of information about skills and experience of employees [31]. Pairs certified, usual ESN contain reliable information on expertise of members.

Globalization of economy has contributed to the crisis in developed countries. To face this problem we need more than reports and databases.

3 Economic and Environmental Context

The current economic situation in the developed countries and intensive industrialization in Asia generate new problems and needs – among them we can mention the industrial decline and unemployment in developed countries, exodus from regions to towns and the emergency of planet protection. In search of the cheapest work force China has become the world factory. Goods travel all around the globe, increasing pollution and amplifying global warming. Asian people are also studying abroad to improve their intellectual capital and sometimes bring it back to their respective countries. However we can also observe growing Diasporas that may influence their origin and adopted countries IC.

In Europe the emphasis is on education and innovation. They are seen as a magic wand to renew industry, impulse growth and job creation. Despite the recommendations of the Lisbon treaty, the impact of education and innovation is still not measured in term of job creation and economic development of the cities, regions and countries. This problem is complex, but with the help of talented people and advanced technology we have in Europe it can be easily solved.

The intensive industrialization from the beginning of 20th century did not taken into account the impact of these activities on livings and the planet - Lenkowa [15], Eckholm [16]. The recent alerts points out the extreme emergency – Arthus-Bertrand [17]. The Earth Summit was launched in 1992 in Rio de Janeiro. Since many others summits discussed the facts, sign agreements, but no action plans are made to change this situation. Business focus society does not change. The Sustainable Development and Corporate Social Responsibility movements focus, among others, on the optimized use of local resources [32].

While companies say to be concerned about carbon and recently about water footprint, less about raw materials, they seem not concern by biodiversity. In reality they still do not manage the human capital; the local skills and know-how are not considered, because of the wrong focus and lack of holistic approach. By consequence skilled people manage themselves and travel to the places their talents are recognized. Despite ubiquitous information and communication technology these movements remain significant even increase.

Public and private organizations have been produced a lot of data using "data thinking" [11]. Data is stored in datacenters producing heat that should be managed; some efforts are now made to explore it [32]. Google and Facebook install them in cold areas (Sweden, Finland in Europe), but despite the efforts to manage the heat system, they influence local ecosystems.

The current trend is to explore data a posteriori and got revenue from the related services. While Google masters exploration of "big data" created on voluntary basis by its users, the other owners of Big Data and Open Data are just thinking about what kind of paid services they can offer. Data mining techniques are available, but mostly statistics are used; however other Knowledge Discovery techniques and tools with AI inside are available since over thirty years now. Knowledge discovery from text may help finding experts, by cross analysis of their written work. Main barrier of effective exploration is the way of thinking and separation between areas.

Technology and in particular AI has a great potential to master the impacts, however these techniques are underused.

The phenomenon of social innovation is a step forward exploring available resources. Internet and mobile applications offer services connecting people having special capacity to offer with those who need this kind of services. Business models of social innovation are just emerging.

The appropriate management of human capital and the education of knowledge cultivators will certainly bring a contribution to planet ecosystems protection.

This challenge is among the most important of the 21 century. It is vital to understand what we have locally and what we need. The intellectual capital should be managed in connection with others tangible and intangible assets of companies and organizations, of cities, of regions and countries using a combination of holistic and system approaches – Mercier-Laurent [11].

As the basic component of Intellectual Capital is Human Capital, we focus on its smart management.

4 Managing Human Capital

While some thinkers state that the human capital is the most important asset, only few are measuring and managing it. The most important barriers in managing intellectual capital is the lack of focus, following trends, lobbying but also ignorance, selfishness, egocentrism, wrong focus and the way of thinking. Internet offered an easy way of producing data and information and amplified one way communication. Increasing number of various services users produce books, articles, pictures and videos, without listening to the others.

The are a plethora of various databases and "big data", built using traditional information processing methods, as a very limited number codes for professions[1], taking into account only traditional ones. Pole d'Emploi (national center for employment) and other public initiatives in France are supposed to help people in finding jobs, but their efficiency is very low, because they are not using the appropriate methods and tools.

Some methods for measuring human capital are available, such as Mediolanum Asset Management[2] method or those of Sveiby [2]; they are not largely used.

With the quick progress of technology and artificial intelligence, computers are able to process natural language instead of codes. This open programming approach allows including in real time new professions that appear every day.

There are also a lot of valuable papers and electronic reports about intellectual capital containing key information and complex charts. They are very useful to know the current status, but they are static, rarely updated and can not be used for supporting a dynamic process of IC management.

For someone looking for a local know-how, it is not easy to find quickly a right person. Some social networks such as Viadeo in France or LinkedIn are trying to connect talents and those who are looking for. Google is certainly among the most efficient search engines, but its business model introduces an important "noise" (and intellectual pollution) due to the advertisement system management.

Human capital may be represented and managed using advanced technology.

One of the first applications of competency management was the hybrid system for long-term and short-term capacity management of French police (Gendarmerie Mobile).

4.1 Managing Gendarmerie Mobile Workforce

Developed in 1989, this application was conceived to allow the short-term management and long-term planning of human resources.

For over 25 years this application is still operational and innovative compared to current methods for managing human capital. The intellectual approach is those of

---

[1] for example all information services are coded 721Z.

[2] http://www.maml.ie/.

"knowledge thinking", while most of human capital managers use "data thinking" and scoreboards as decision support.

The Mobile Gendarmerie (in French: *Gendarmerie Mobile*) is a subdivision of the French Gendarmerie. Specific anti-riot units of the Gendarmerie were established at the beginning of the 19th century. The name of Gendarmerie Mobile (GM) was given in 1921.

Mobile gendarmerie[3] is a workforce of 17,000 men who ensure the missions of maintaining and restoring order and operational defence of the territory. Their missions are following:

- Riot control and order recovering
- Monitoring the territory and its dependencies
- Guarding sensitive locations such as embassies, airports, railways stations, etc.
- Defending the territory, providing reinforcements to the departmental Gendarmerie
- Conducting external operations

The management of GM workforce consists of:

- Allocation of the appropriate teams for a given mission
- Know and follow the activity of all units (groups and squadrons).

There are permanent missions, such as custody of the Embassy and specific ones, such as crisis management. The permanent missions are planned in advance and change periodically, against the specific missions require management to task - it comes to identifying qualified people and located near of intervention place to perform actions.

The architecture of developed solution is presented in Fig. 1.

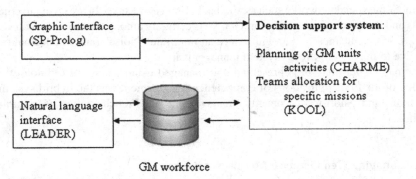

GM workforce

Fig. 1. ARAMIS system for Management of skills and know-how of Mobile Gendarmerie [3].

The hybrid architecture of ARAMIS decision support system is composed of following modules:

---

[3] http://www.gendarmerie.interieur.gouv.fr/fre/Notre-Institution/Nos-composantes/Gendarmerie-mobile.

- Graphic interface written in SP-Prolog
- Database contains information on police skills, know-how and experience, their location, the tasks completed and in progress and planned leave as well.
- Support system for planning ongoing activity of units. The permanent missions module implemented in constraint programming language CHARME[4], takes into consideration the skills needed to perform a given task, precedence constraints (conducted missions and locations) and the number of units required. The use of CHARME is justified by the combinatorial character of this problem: over 100 squadrons split into groups, more than 100 types of missions, each lasting between 15 and 120 days, more than 20 features to consider, eight months of activity to plan. The system allows simulations to achieve optimal planning and quick rescheduling by performing minimal changes.
- Expert systems (KOOL[5]) for short-term resource allocation for specific missions such as the visit of a foreign president or ORSEC plan[6] deals with the characteristics of the mission to be performed, the ability to perform a given task, the availability of units and verifies the proximity of these groups. The allocation process requires knowledge of domestic regulations specific for the mobile gendarmerie. An example of rule:

If there are no units available, and it is a vital and urgent mission (life-threatening, major damage), close and short duration, *then* allocate the team currently being trained.

The expertise of this project was reused for another resource allocation system for the Winter Olympic Games (OG) in Albertville, 1992.

4.2   Resources for Security Management During the Olympic Games

For the needs of the Olympic Games in Albertville, 1992 the above described system was extended and integrated to overall OG management systems, presented in Fig. 2. The main objective was the optimal resources allocation for permanent missions a well as for specific missions in crisis situations, such as avalanche, transport accident, blocked road etc.

Context. 1600 km^2 of high mountain area, thirteen Olympic sites, over one million of spectators, 1200 tourism coaches and 150 helicopters. Risks to manage are those of High Mountain and of accidents du to significant road-rail-air traffic, crowd gathering.

Specifications. The system must be able to handle multiple crises simultaneously and allow the rapid response. Each site must be equipped with decision support system. A central command-control system must handle all crises even outside the local level. The system must manage the allocation of all types of security units, such as Police,

---

[4] CHARME - the first constraint programming language developed by Bull CEDIAG from the research results of the European Computer Research Center (ECRC), Munich.

[5] KOOL – Knowledge representation Object Oriented Language, developed by Bull CEDIAG.

[6] ORSEC plan - French generic emergency plan in case of disaster, when the local means are not sufficient.

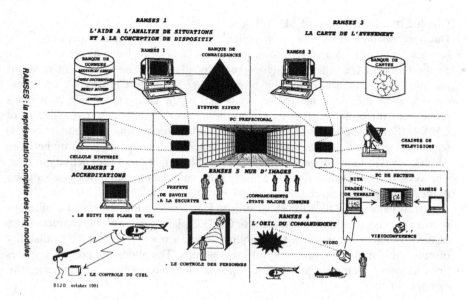

Fig. 2. Ramses – Integration to existing command and security control system [4]

Firefighters, Chasseurs Alpins (the elite of mountain infantry of the French Army), CRS (*Compagnies républicaines de sécurité*[7]), and others, in record time and in sufficient quantity.

System Architecture. This system, built up from the expertise of more than 60 specialists, supports the decision maker in analyzing the situation and composing appropriate and available resources and teams for intervention.

The RAMSES 1 expert system, shown in Fig. 2, analyzes the situation; suggest the actions and the all types of resources needed. The topographic display module locates the close resources available. In the case of necessity to manage simultaneously multiple crises the Optimization Module chooses optimal solution.

These three modules interact with the database including information on the human and other resources, traffic, weather conditions, accommodations, risky locations, the directory of civil security, and others.

The Olympic area was split into seven sectors, each placed under the responsibility of a sub-prefect controlling the overall security system. The Central Command Post take decisions when local resources are insufficient.

Such a decision support system may be useful in many cases.

Another way of managing talents is simply to know them. The easiest way is to display a content of human resource database. Sur, to provide a real help, it should be up-to-date and updated in real time.

---

[7] Republican Security Companies.

4.3 Know What We Have

A concept of "knowledge trees" have been introduced by Autier and Levy [18] and implemented in tools as Ginko (Trivium) and Selva (Ligamen), offering graphic representations of individual and collective skills as a tree.

The Fig. 3 illustrates the skills of 10 people: the trunk represents common knowledge, branches the specialties, and leaves the unique skills. Such an image provides information on everyone's ability and helps to decide if the unique skills represented by the leaves are strategic.

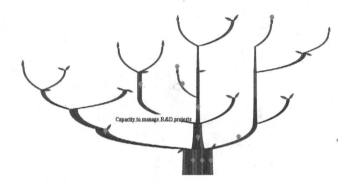

Fig. 3. "Knowledge tree" created using the Ligamen software (http://www.ligamen.fr)

The part to the far right of the trunk, as well as the triple branch, indicate the position of a person in a group. Such visualization facilitates the identification of skills and helps to detect the lacks in relation to a required profile, which can be filled by training. Thus, we can build the competency tree of a company, city or a region and reason backwards: what projects can we achieve with such intellectual capital? We then need to search for the skills in a neighbouring region or "rent" them to a partner.

In the international context and within a networked enterprise, it would be better to manage skills with a holistic perspective - on regional, national and international levels. This intangible wealth can grow through continuous learning from interaction with the environment, according to corporate strategy. However it is necessary to define some rules for information updating and validation – who can/should update or who validate this update? Is the up-date allowed in real time or periodically?

The training department is in charge of making this capital grow and generating value from it. It is involved for now, because in the global Knowledge Management approach all knowledge cultivators are constantly learning things in function of current needs and their own ambitions.

The training department could also manage the transmission and preservation of the essential knowledge and know-how of retiring, especially when this is the knowledge of a long-life and a strategic product for the company. Collaboration between several professionals facilitates the skills management.

4.4   How to Find the Right Profile

When we know what we have and what we are looking for, one of artificial intelligence techniques - case-based reasoning [19] could be very helpful. The built-in analogy engine works by matching demand (*I am looking for*) and offer (*base of existing skills and know-how*) to find instantaneously the profile we want, if such a profile is registered. If not, a set of similar profiles that could be adapted to the expected ones by training is proposed to the user. We can imagine a World Knowledge Base including Talent Bank equipped with such an engine. The principle is presented in Fig. 4.

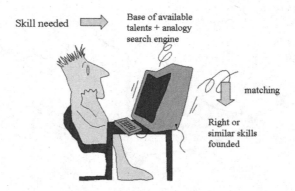

Fig. 4. Principle of case-based reasoning applied to matching offer and demand [28]

The user asks for a given profile using appropriate interface, the analogy engine retrieve from the talent bank the right profile if it exists or close to what is asked for. The user can decide the best person having the needed competency. If he/she decide to hire someone with similar profile to demand, this person can be trained to improve his/her capacity; the case base should be updated to serve for future matching.

Each company, organization, city, region and country can use such a tool for smart competency management.

5   From Talent Management to Organizational Strategy

The various methods for measuring the value of human capital of a company, city, region or country may provide the information on what we have. The same information can be explored to plan the future development - what new activities and companies may be created from existing IC. This purpose is illustrated in Fig. 5.

A company/city/region need to elaborate a clear vision for the future. At this point the skilled persons able to envision it are needed. This vision will be "translated" into corporate (organizational) strategy and a tactics (actions to achieve the strategic goals). The intellectual capital of professional working on accomplishing various tasks grown and the new knowledge and capacity should be taken into account at strategic level. It may also influence the vision.

Fig. 5. Human capital dynamics

To go further, it is vital to consider the relations between the smart IC management and the sustainable success of a given organization. It involves several ecosystems that can be balanced to ensure such a success. An example of such ecosystems is shown in Fig. 6.

Fig. 6. Ecosystems of human capital

Current organizational strategy is decided at top management level and considers HR just as a workforce serving to achieve the goals. The question: How we can innovate better exploring our talents?" is not asked. Educational programs are defined by Ministry of Education without taking into consideration the future market needs in term of skills. There are many new skills needed and many are obsolete. The communication and knowledge of each component of these ecosystems is a vital element for preserving the balance, be successful and avoid unemployment.

6   Skills for the Future

Today educational system produces the traditional professionals. Many of them face the difficulty in finding job in their region or country. The most audacious travel for job; change country, language and continent. They have to adapt to new conditions and to new culture.

As mentioned before, our future depends on our capacity to adapt, to detect opportunities, to collect necessary skills and knowledge and to transform them into economic values, in balance with ecosystems. It also depends on the rapidity of our decision making, on our risk taking ability in a dynamic environment, and on our ability to use the computer, regardless of its form, as an intelligent assistant. The latter facilitates an innovation without boundaries between fields (out of the box thinking).

Facing the affluence of information and solicitations, a new skill is required – the innovation know-how. This is the art of finding and exploiting strategic information and of gathering momentum and developing the knowledge and skills essential to the success of this enterprise, which is innovation in its entirety.

These skills are numerous – from the management of ideas and people to the implementation and commercialization [11].

Although Europe has a long innovation tradition since the industrial era, globalization has changed the odds. The factors such as the slowing down, the obsolescence of some sectors and the emergence of others, as well as the relocation of activities, influence active knowledge and skills. The lack of interest expressed by youths in scientific studies will lead to a shortage of engineers. Some skills are disappearing with retirements, which are sometimes accelerated by the economic crisis. Knowledge capitalization approaches are saving a part of the strategic and "sensible" skills, but these initiatives are quite rare and are often initialized too late.

The European document Putting Knowledge into Practice (European Commission 2004) specifies that the lack of skills, notably in the fields of sciences, engineering and ICT, is a challenge for European education. Another publication, Innovate for a Competitive Europe (European Commission 2004) advises companies to learn how to transform the absorbed knowledge into action. Such an innovation dynamics combines

Fig. 7. European Union vision of the skills for Future [29]

the knowledge and skills in value creation. Kolding et al. [22] describe the skills we need to acquire to face the post crisis era in Europe. The authors are convinced that the ICT skills are the most important, but they did not mention what approach to ICT and to computer programming should be used. Figure 7 shows the EU vision of ICT skills needed for the future.

Is it really enough to develop just ICT skills to build a successful Europe?

Companies training departments need to focus on the transformation of today capacities good for industrial economy to those that are essential for Knowledge Economy. The progress can be measured using for example the trees of knowledge software, or other that may help.

7   Conclusion and Future Works

Some approaches presented above may seem old, but apparently there are no new intellectual approaches for human capital decision support systems. All available works in this area describe "Intellectual Capital Reports", mainly paper, database and scoreboards. The users of Organizational Social Networks consider themselves innovative. Consequently, human capital professionals, as well in companies as in public organizations, still do not have efficient tools for measuring, growing and effective using of this capital.

To build a sustainable future we need more than data base, reports and dashboards, we need a disruptive innovation in the way we build, evolve, maintain and manage the human capital.

We need a new educational system, having the ambitious task of changing mentalities and values, to educate a culture of knowledge cultivators and to increase imagination and creativity. Main challenge of education is to teach how to learn, the curiosity, adaptability, capacity of solving problem with limited resources and to undertake and succeed collectively. This education is based on exchanges, listening and respecting the others opinions; an education for all, to learn from nature, from the past and from differences, in which technology and means of communication have a significant role to play.

We need to use the power and "intelligence" of computers and other connected devices. When programmed using "knowledge thinking", they can bring a significant helps in storing, updating, displaying, matching and finding the relevant elements of human capital.

Computers in all forms should be programmed to work in synergy with their users in am to help them in the tasks computers are better than humans. The same logic needs to be applied for robots and drones design, for connected object and internet of things. Games have a special role to play in long-life education – to provide knowledge by playing instead of having boring lectures.

We need to create synergy between educational programs and local needs and link them to a dynamics vision for the future.

New metrics could be: boldness, imagination, associations (links making), and capacity to find and use the appropriate knowledge, mental flexibility, knowledge and ecosystem thinking, capacity to transform ideas in value and to envision the future. The estimation of 5D impacts of resulting activities – economic, technologic, cultural, social and environmental, could be added to measure the progress.

Such a wise management of intellectual capital, supported by electronic "intelligent" assistants and appropriate measure of progress is essential for the development of companies, regions and countries.

References

1. Schultz, T.: Investment in human capital. Am. Econ. Rev. **51**(1), 1–17 (1961)
2. Sveiby, K.E.: The New Organisational Wealth - Managing and Measuring Knowledge -Based Assets. Berrett-Koehler, San Francisco (1997)
3. Geraud, N., Rincel, P., Vandois, N.: ARAMIS-GM Un système intelligent d'aide à la décision pour la gestion des effectifs de Gendarmerie Mobile, Systèmes Experts et leurs applications, Avignon (1990)
4. Lacroix, V.: Lieutenant Colonel Daville: RAMSES I in système d'aide à la décision pour la sécurité des Jeux Olympiques, Systèmes Experts et leurs applications, Avignon (1991)
5. OCDE 1996, Measuring What People Know. Human Capital Accounting for the Knowledge Economy (1996)
6. Becker, G.S.: Human Capital: A Theoretical and Empirical Analysis, with Special Reference to Education. University of Chicago Press, Chicago (1964). ISBN: 978-0-226-04120-9
7. Edvinsson, L., Malone, M.S.: Intellectual Capital: Realizing your Company's True Value by Finding Its Hidden Roots. Harper Business, New York (1997)
8. OECD, The Well-being of Nations. The Role of Human and Social Capital. Education and Skills (2011). http://www.oecd.org/site/worldforum/33703702.pdf
9. Dixon, P., Gorecki, J.: Sustainagility. How Smart Innovation and Agile Companies will Help Protect our Future, p. 232, 20. Kogan Page Publishers, London (2010)
10. Berkes, F., Colding, J., Folke, C. (eds): Navigating Social-Ecological Systems, pp. 352–387. Cambridge University Press, UK (2003)
11. Mercier-Laurent, E.: Innovation Ecosystems, p. 248 Wiley (2011). ISBN: 978-1-84821-352-8
12. Aubin, J.P.: Viability Theory. Birkhauser, Boston (1991)
13. Savage, C.: 5th Generation Management: Integrating Enterprises through Human Networking. The Digital Press, Bedford (1990)
14. Amidon, D.: The Innovation Strategy for the Knowledge Economy. Heineman Butterworth, Boston (1997)
15. Lenkowa A.: Oskalpowana ziemia, Omega, Wiedza Powszechna, Warsaw, Poland (1969)
16. Eckholm, E.P.: Losing Ground. Environmental Stress and World Food Prospects. W.W. Norton and Company, New York (1976)
17. Arthus-Bertrand, Y.: Home (2009). https://www.youtube.com/watch?v=jqxENMKaeCU
18. Autier, M., Lévy, P.: Les arbres de connaissances. La Découverte, Paris (1992)
19. Kolodner J.: Case-Based Reasoning, p. 668. Morgan Kaufman (1993). ISBN: 978-1558602373
20. European Commission, Implementing the partnership for growth and jobs: Making Europe a pole of excellence on corporate social responsibility (2006)
21. European Commission, Innovate for a Competitive Europe. A New Action Plan for Innovation, 2 April 2004

22. Kolding, M., Ahorlu, M., Robinson, C.: Post crisis: e-skills are needed to drive Europe's innovation society, IDC EMEA, London, UK (2009)
23. Youriev, A.M.: History of human capital (2014). http://www.yuriev.spb.ru/polit-chelovek/human-capital-resource
24. Skyrme, J.D.: The Seven Ages of Information and Knowledge Management. http://www.skyrme.com/kmarticles/7ikm.pdf
25. Ordonez de Pablos, P., Edvinsson, L. (eds): Intellectual Capital in Organizations. Non Financial Reports and Accounts, p. 316. Routledge, Taylor & Francis Group (2015). ISBN: 978-0-415-73782
26. Lin, C.Y.Y., Edvinsson, L.: National intellectual capital model and measurement. Int. J. Knowl. Based Dev. 3(1), 58–82 (2012)
27. Stahle, P., Stahle, S., Lin, C.Y.Y.: Intangibles and national economic wealth: a new perspective on how they are linked. Int. J. Intellect. Capital, Emerald Insight 16(1), 20–57 (2015)
28. de Mantaras, R.L., Plaza, E.: Case-based reasoning: an overview. AI Commun. 10(1), 21–29 (1997)
29. Horizon 2020 Work Programme 2014–2015, Europe in Changing world; inclusive, innovative and reflective societies, European Commission C(2013), December 2013
30. Wu, A.: The integration between Balanced Scorecard and intellectual capital. J. Intellect. Capital 6(2), 267–284 (2005)
31. Le Moing, B.: Schneider-Electric: La gestion des connaissances au cœur du programme d'entreprise, Qualitique, No. 256, pp. 46–48, Novembre 2014
32. Allais, R., Reyes, T., Roucoules, L.: Inclusion of territorial resources in the product development process. J. Cleaner Prod. (2015)
33. Mercier-Laurent, E.: The Innovation Biosphere – Planet and Brains in Digital Era. Wiley (2015)

Author Index

Ahmed, Rajib 107
Alsqour, Moh'd 21
Atifi, Hassan 43

Bhuiyan, Touhid 107

Conruyt, Noël 1

Duzdar, Irem 150

Grosser, David 1

Hauke, Krzysztof 92

Kalisch, Mateusz 132
Kayakutlu, Gulgun 150

Matta, Nada 43
Mercier-Laurent, Eunika 150, 165

Owoc, Mieczysław L. 21, 92

Perechuda, Kazimierz 58
Pondel, Maciej 92
Przystałka, Piotr 132

Rauscher, François 43
Razzak, Mohammad Abdur 107
Rousseaux, Francis 75

Saurel, Pierre 75
Sébastien, Didier 1
Sébastien, Olivier 1
Sébastien, Véronique 1
Sennaroglu, Bahar 150
Sobińska, Małgorzata 58

Timofiejczuk, Anna 132